Complementarity Beyond Physics

Arun Bala

Complementarity Beyond Physics

Niels Bohr's Parallels

palgrave
macmillan

Arun Bala

ISBN 978-3-319-39783-2 ISBN 978-3-319-39784-9 (eBook)
DOI 10.1007/978-3-319-39784-9

Library of Congress Control Number: 2016947004

Cover icon created by EliRatus from the Noun Project

Printed on acid-free paper

This Palgrave Macmillan imprint is published by Springer Nature
The registered company is Springer International Publishing AG Switzerland

Acknowledgements

I would like to acknowledge Asia Research Institute (National University of Singapore), Rotman Institute of Philosophy (Western University), and Department of Philosophy (University of Toronto) for offering me positions, facilities, and an atmosphere of indispensable intellectual stimulation that made this study possible. I would also like to tender my heartfelt gratitude to many friends and colleagues who helped to enhance both the range and rigor of this work. In particular, I would like to thank James Robert Brown, Andrew Brennan, Donald Wiebe, Prasenjit Duara, Richard Arthur, Lee Keekok, Joergen Moeller, Ng Chee Yuen, Wayne Myrvold, and Ten Chin Liew for many stimulating discussions and invaluable feedback. I would also like to express my indebtedness to my teachers Anthony Leggett and Jeffrey Bub who first set me probing more deeply the philosophical issues raised by the quantum theory, and led me to wonder whether Bohr was right after all. Finally I would like to voice my deepest gratitude to my partner-in-life Selvarani for her patience, encouragement, and unrelenting support.

Contents

List of Figures

CHAPTER 1

Complementarity Beyond Physics

1.1 Niels Bohr and Eastern Thought

In his book, *Atomic Physics and Human Knowledge*, the physicist Niels Bohr suggests that in order to deepen our understanding of the philosophical problems generated by the quantum theory we should turn to ancient Indian and Chinese thinkers. He writes:

> For a parallel to the lesson of atomic theory … [we must turn] to that kind of epistemological problems with which already thinkers like Buddha and Lao Tse have been confronted, when trying to harmonize our position as spectators and actors in the great drama of existence. Still, the recognition of an analogy in the purely logical character of the problems which present themselves in so widely separated fields of human interest does in no way imply acceptance in atomic physics of any mysticism foreign to the true spirit of science, but on the contrary it gives us an incitation to examine whether the straightforward solution of the unexpected paradoxes met with in the application of our simplest concepts to atomic phenomena might not help us to clarify conceptual difficulties in other domains of experience. (Bohr 1958: 19–20)

The passage above makes evident that what motivates Bohr to turn to Eastern thinkers is not a religiously inspired mystical impulse but his belief that their views closely parallel the apparently paradoxical epistemological ideas in the complementarity framework he formulated to interpret quantum theory.[1] This framework assumes that we need to use mutually exclusive concepts such

Arun Bala, *Complementarity Beyond Physics*,
DOI 10.1007/978-3-319-39784-9_1

as particle and wave in dealing with atomic entities. For example, in classical physics if an entity were taken to be a particle it was assumed that it could not be a wave at the same time. Thus, at the end of the nineteenth century, in classical physics it was assumed that electrical charges were carried by particle-like entities such as electrons and that light was an electromagnetic wave. Since a particle has a specific location in space and a wave is an entity distributed over space it was assumed in classical physics that an entity could not exhibit both particle-like and wave-like behavior. However, with quantum theory it became evident that light also exhibited particle-like behavior so that it could be seen as a stream of photons, and that electrons also could behave like waves under certain experimental situations. Thus an electron or a photon could respond as a particle or a wave under different experimental situations. Although it might seem that 'particle' and 'wave' are mutually exclusive concepts, quantum theory requires us to assume that they are complementary descriptions of entities in the micro-world.

In this context the concept of mutual exclusivity must be recognized to mean not that we cannot include the concepts in one theoretical framework by attributing them to the same object, but that we cannot visualize in one picture how something can be both a particle with a definite location in space and time and a wave dispersed over space and time. Nevertheless we cannot avoid using such classical concepts in order to both describe our experimental set-up and to present the results of our observations. It is their mutual exclusivity vis-a-vis visualization and equal necessity vis-a-vis conceptualization for understanding quantum processes that leads us to treat particle and wave aspects of phenomena as complementary. Bohr has emphasized this point:

> The extent to which ordinary physical pictures fail in accounting for atomic phenomena is strikingly illustrated by the well-known dilemma concerning the corpuscular and wave properties of material particles as well as of electromagnetic radiation. It is further important to realize that any determination of Planck's constant rests upon the comparison between aspects of the phenomena which can be described only by means of pictures not combinable on the basis of classical physical theories. ... In this situation, we are faced with the necessity of a radical revision of the foundation for description and explanation of physical phenomena. Here, it must above all be recognized that, however far quantum effects transcend the scope of classical physical analysis, the account of the experimental arrangement and the record of the observations must always be expressed in common language supplemented with the terminology of classical physics. (Bohr 1950: 51–52)

However, there is another dimension to Bohr's claim that ordinary physical pictures cannot account for atomic phenomena. This arises because Bohr also argues that objects do not have properties such as position and momentum intrinsically and independent of their determination by a measuring device. They arise within the context of a measurement although the system itself is defined in terms of these properties prior to its measurement.[2] We may say that the system is defined in terms of properties that it can exhibit under different, mutually exclusive, measurement contexts. Although the notions of a precise momentum and a precise position can be combined into a single picture in classical physics, quantum theory precludes this. They have to be seen as mutually exclusive, but complementary, attributes of atomic entities.

The simultaneous use of mutually exclusive concepts as complementary would appear to violate the laws of logic generally associated with classical science. These were formulated very early by Aristotle in terms of three laws—the law of identity, the law of noncontradiction and the law of the excluded middle. The law of identity states "that everything is the same with itself and different from another". The law of noncontradiction states, in the words of Aristotle, that "one cannot say of something that it is and that it is not in the same respect and at the same time." The law of the excluded middle states that something either has a property P, or does not have the property P.[3] However, complementarity seems to violate these laws of Aristotelian logic—a point noted by the eminent theoretical physicist, Robert Oppenheimer, who writes:

> If we ask, for instance, whether the position of the electron remains the same, we must say "no"; if we ask whether the electron's position changes with time, we must say "no"; if we ask whether the electron is at rest, we must say "no"; if we ask whether it is in motion, we must say "no." (Oppenheimer 1954: 40)

He adds that the Buddha had given similar answers when interrogated as to the existence of a person's self after death although such answers are unfamiliar to the tradition of seventeenth and eighteenth century science.

Oppenheimer's observation suggests that not only does quantum theory seem to violate the rules of classical logic but also that there are traditions of Eastern thought which have come to recognize, long before quantum theory, that to comprehensively understand phenomena in the universe often requires the use of mutually exclusive but complementary notions.

Thus in Madhyamika Buddhism there developed a tradition of logic, closely linked with its founding philosopher Nagarjuna (c.150–c.250CE), which suggested that we could impute predicates to an object that would suggest it has a property P, does not have P, has both P and not-P, or has neither P nor not-P. Known as the logic of the *catuskoti*, it has parallels to the Greek tetralemma associated with the skeptics. However, in the West the tetralemma came to be a part of a marginalized tradition, while the logic of the *catuskoti* deeply influenced not only logical traditions in India but also those in China and East Asia as a result of the wide influence of the Madhyamika tradition.[4]

Moreover, in China there were also similar paradoxical positions held by the Daoists. These were formulated in a slightly different way from the *catuskoti*, although they also implied violations of the laws of classical logic. This can be seen in paradoxes they formulated such as 'the action of no-action', 'the knowledge of no-knowledge', and 'the morality of no-morality.'[5] There is a striking affinity in this style of thinking with the notion in quantum physics that something could be both a particle and yet not a particle by virtue of being a wave. It is these apparent violations of normal logic that motivates Bohr to propose that we could profit from a dialogue with Eastern, especially Buddhist and Daoist, traditions of epistemologies.

Bohr is not alone in recognizing that dialogue with Eastern traditions can illuminate the epistemological implications of quantum physics. Werner Heisenberg, who also played an equally seminal role in the birth and development of quantum theory and cooperated with Bohr in formulating the Copenhagen Interpretation of quantum theory that has dominated the way physicists think about the atomic world even today (albeit with diminishing force following other interpretations such as the many-worlds and hidden variables interpretations), makes the same point:

> [T]he great scientific contribution in theoretical physics that has come from Japan since the last war may be an indication of a certain relationship between philosophical ideas in the tradition of the Far East and the philosophical substance of quantum theory. It may be easier to adapt oneself to the quantum-theoretical concept of reality when one has not gone through the naive materialistic way of thinking that still prevailed in Europe in the first decades of this century. (Heisenberg 1963: 173)

Japan is of course the land of Zen philosophy which in many ways is the synthesis of Daoist and Buddhist thought. It originated in China as *Chan* during the Tang dynasty (618–907), but its distinctive style of Buddhism spread to Vietnam in the late sixth century, Korea in the seventh century, and finally Japan in the twelfth century, where it came to be labeled Zen (Dumoulin 1988, 1990).

Hence it is not surprising when, in response to the question raised by the physicist Leon Rosenfeld—one of Bohr's close collaborators and friend—whether Japanese physicists had any difficulty understanding Bohr's views on complementarity, Hideki Yukawa, the Japanese Nobel laureate in 1961, answered:

> No, Bohr's argumentation has always appeared quite evident to us; … You see, we in Japan have not been corrupted by Aristotle. (Quoted in Rosenfeld 1963: 47)

Rosenfeld adds "Bohr rediscovered the dialectical process of cognition which had so long been obscured by the unilateral development of epistemology on the basis of Aristotelian logic and Platonic idealism." One can surmise that Yukawa is likely to have been motivated by the recognition that in treating something as both a wave and a particle, or as neither a wave nor a particle, we seem to be violating two of the three important laws of thought that underpin standard logic as formulated by Aristotle.[6] Indeed complementarity seems to suggest that we need to treat Aristotelian logic as an empirical discovery refuted by quantum theory in the same sense that general relativity theory suggested transcending Euclidean geometry. Such a move seems to have inspired the philosophers Hilary Putnam and Michael Dummett who, following the work of Garrett Birkhoff and John von Neumann on quantum logic, were led to argue that logic itself should be treated as an empirical science.[7]

In addition to noting the parallels to complementarity in Eastern thought, Bohr also made sustained and systematic efforts to develop the implications of parallels to complementarity beyond the domain of microphysics. Indeed Bohr used every opportunity over the last three decades of his life to promote his conviction that the epistemological insights inspired by quantum theory must be applied to other areas of science. He writes:

> The epistemological lesson we have received from the new development in physical science, where the problems enable a comparatively concise formulation of principles, may also suggest lines of approach in other domains of knowledge where the situation is of essentially less accessible

> character. An example is offered in biology, where mechanistic and vitalistic arguments are used in a typically complementary manner. In sociology, too, such dialectics may often be useful, particularly in problems confronting us in the study and comparison of human cultures ... Recognition of complementary relationship is not least required in psychology, where the conditions for analysis and synthesis of experience exhibit striking analogy with the situation in atomic physics. (Bohr 1950: 54)[8]

His articles then go on to delineate how the complementarity perspective could be widened to include, in particular, our epistemological understanding of psychological, biological and anthropological processes.[9]

Nevertheless, Bohr's reputation and prestige have not convinced the scientific community that they should take seriously either his efforts to engage Eastern thought or his attempts to extend complementarity into other scientific disciplines. This is underlined by Henry Folse in his book, *The Philosophy of Niels Bohr* written more than 20 years after Bohr's demise:

> Bohr had envisioned complementarity spreading into wider and wider fields, just as the mechanical approach of Galileo had started in astronomy and simple phenomena of motion and gradually spread to all of the physical sciences. However, thus far history has proved him unduly optimistic, for instead physicists and philosophers have concentrated on more and more detailed problems and have developed increasingly sophisticated analyses of the theoretical formalism. (Folse 1985: 168)

Why have Bohr's wider epistemological endeavors fallen on deaf ears despite his indisputable standing as a scientist? Two factors can explain this neglect of Bohr's wider project for complementarity. First, even in the early years of the quantum revolution there arose strong opposition to quantum theory spear-headed by Albert Einstein and including other highly regarded physicists such as Erwin Schrodinger and Max Planck. These esteemed members of the physics community, despite having made significant contributions that led to quantum theory, saw the theory as incomplete because it failed to give a causal and deterministic account of atomic phenomena. They argued that Bohr's epistemology of complementarity, with its appeal to mutually exclusive concepts, violated both logic and reason and masked the limitations and inadequacies of quantum theory. Such reservations expressed by prominent colleagues made complementarity questionable even within the domain of physics and served to discourage scientists from following Bohr's attempts to pursue its epistemological implications into other disciplines beyond physics.[10]

Second, the way Bohr in the earlier years explained the need to use complementary notions in the atomic domain led many scientists, even those conceding its relevance for physics, to infer that it could not have applications beyond the atomic realm. Bohr argued that complementary concepts were required in the atomic domain because of the uncontrollable interactions that occurred whenever a scientist deployed a physical apparatus to measure the properties of a micro-system. For example, to determine an electron's position scientists have to set up a position measuring apparatus with which it is made to interact. But this interaction itself disturbs the electron in a way that cannot be controlled so that we lose precise information about its momentum. Similarly if we wish to have more precise information about its momentum we have to set up a momentum measuring apparatus, but our inability to prevent uncontrollable interactions with it would lead to loss of information about its precise position. Bohr argued that such disturbances produced by the scientist's act of measurement occur in all systems but are extremely small. Consequently they were safely ignored in classical physics, which dealt with material bodies much larger than atomic systems. By contrast, he maintained, they become extremely significant at the atomic level.

However, by tracing the need for complementary viewpoints to uncontrollable interactions at the micro-level, Bohr seems to implicitly imply that his efforts to apply complementarity to biological, psychological, and anthropological phenomena are not likely to bear fruit. These involve macroscopic systems much larger than quantum micro-systems and more similar in size to those studied by classical physicists which exhibit no complementarity effects. Consequently any 'uncontrollable interactions' with the measuring system can be ignored in their cases because their effects would be negligible. This offers no incentive for scientists to look for ways to extend the applications of complementarity to such macro-systems.

However, it may be argued that although Bohr did base his arguments for complementarity on the 'disturbance of measurement on the system' he gave up this approach to defend complementarity after 1935, following the debates he had with Einstein on the EPR thought experiment. But his reformulation of the argument technically in terms of the notion of 'action' in physics also suggests that complementarity cannot be applicable to macro-systems. The action principle involves reformulating differential equations of motion for physical systems as an equivalent integral equation. There are several variants of this approach but the most common action principle is Hamilton's principle. In its formulation for classical mechanics the principle suggests that the path followed by a physical system will be one where the action is minimized, or strictly speaking is stationary. It allows the classical equations of motion of a system to be derived by minimizing the value of the action integral instead of solving differential equations. Indeed the two approaches are equivalent since Hamilton's principle allows us to formulate the differential equations for any physical system by an equivalent integral equation. The approach has wide applicability for not only does it apply to classical mechanics but also to classical fields such as the electromagnetic and gravitational fields. More significantly it can be extended to quantum mechanics and quantum field theory by making use of the concept that a physical system simultaneously follows all possible paths where the action for a path determines the probability amplitude for it. Indeed coupled with the notion of the quantization of action, it explains the roots of complementarity in quantum physics. However, it also suggests that complementarity can be ignored in the macro world where the effects of quantization of action are extremely small.[11]

The historian of science Mara Beller, in her study *Quantum Dialogue: Making of a Revolution*, uses this extension of the action principle to quantum physics as a powerful argument to repudiate Bohr's efforts to extend complementarity into areas beyond physics:

> The analogies of complementarity are partial, incomplete, vague, and contradictory. This is not surprising for, according to Bohr, the main conclusions of his complementarity philosophy follow from the indivisibility of the quantum of action, yet there is no vigorous analogue of this indivisibility in non-quantum domains. Consequently, no meaningful structural network of analogies can be established between, say, quantum physics and psychology. (Beller 1999: 264)

However, Bohr's complementarity viewpoint has also disturbed scientists for reasons that go beyond questions regarding the adequacy of quantum theory or the applicability of complementarity outside the atomic domain. These scientists are alarmed at the way Bohr's complementarity standpoint has been invoked to support the relativism and antirealism of postmodern social constructivists and the intuitionism and mysticism of New Age thought—the former subverting the notion of scientific realism assumed by most of the scientific community and the latter the equally strong commitment of scientists to the empiricism of the senses. The postmodern critiques of scientific realism are motivated by the recognition that the scientist's choice of observational arrangement crucially determines the properties observed in quantum systems, since whether a quantum system will be seen as having particle-like or wave-like properties depends on which of two different and mutually exclusive measuring arrangements we set up to observe it. This suggests that there are no objectively real properties in the world observed by scientists, but only properties constructed by their social choice of what to observe. By contrast New Age thinkers acknowledge that there are real properties, but argue that the parallels to Eastern thought rooted in mystic experience means that we can acquire knowledge of quantum entities by techniques of intuition that involve internal introspective approaches quite distinct from the empirical use of the external sense organs that deliver knowledge to scientists.

As a result the controversies surrounding complementarity, originally seen as a debate within the community of natural scientists, specifically physicists, has now come to be associated with a much wider conflict—the quarrel between natural scientists defending a heritage of not only scientific realism with postmodern antirealists but also scientific empiricism with New Age mystics.[12] These continuing controversies between natural scientists and their postmodern and New Age critics first came to a head more than two decades ago when they came to be dubbed as "the science wars" because of the violence and intensity of the antipathies they generated at that time.[13] Significantly they led many scientists to conclude that Bohr's epistemology of complementarity had untoward wider implications, which undermined regard for scientific realism and empiricism and, therefore, had to be replaced by philosophically more palatable interpretations of quantum theory.[14] Consequently, although Bohr's scientific views are still highly regarded by practising physicists, there has also been an increasing tendency to see his philosophical standpoint centered on the so-called Copenhagen interpretation as problematic, and a consequent shift to other interpretations, especially the many-worlds interpretation, or to lose interest in interpretations altogether.

Strangely, but perhaps not surprisingly, as many natural scientists came to repudiate Bohr's philosophical views, some social scientists embraced it. The philosopher-sociologist Steve Fuller argues that Bohr implicitly supports the postmodern turn connecting science and society. In his book, *The Philosophy of Science and Technology Studies*, Fuller points out that some of the pioneers of quantum theory not only anticipated current postmodern views but also drew attention to the striking parallels noticed by Bohr between the complementarity principle and the metaphysics of Daoism and Buddhism. These observations lead Fuller to charge that modern day scientists, who criticize Bohr's views, are 'neo-Puritans' vainly attempting to separate technical (i.e. scientific) and cultural matters in a way that quantum complementarity and postmodern theory have shown to be futile.[15]

Remarkably both sides in the science wars came to the conclusion that Bohr's complementarity perspective subverts scientific realism and scientific empiricism and overlook the possibility that Bohr may actually be offering a third epistemological alternative quite distinct from either of their two positions, with radically new conceptions of scientific realism and empiricism that cut across the epistemological divide separating the contestants in the science wars. Indeed we will find that this is the case, and that it provides good grounds for extending complementarity, as Bohr recommends, into areas of knowledge such as biology, psychology, and anthropology, as well as opening a dialogue with Eastern philosophical traditions.

1.2 New Age Appropriations of Complementarity

Let us begin with the Eastern traditions. What is noteworthy is that there is a general disinclination on the part of natural scientists to pursue Bohr's interest in Chinese and Indian philosophies as resources to illuminate epistemological concerns in science. Indeed academic aversion to opening a dialogue with Eastern thought is even greater than the reluctance to follow Bohr's efforts to extend complementarity into areas of knowledge such as biology, psychology, and the social sciences. This explains why there is a much larger body of philosophical and historical literature which discusses Bohr's application of complementarity to the sciences in general rather than his views concerning its parallels in Eastern philosophy.[16] One major explanation for this reluctance to pursue the parallels to Eastern thought is the fear of linking quantum physics, one of the greatest achievements of science, with Eastern philosophies perceived to be closely linked to Eastern religions. Given the historical struggle of science against religious dogmatism in the modern era—a struggle that has not abated even today—such reservations are understandable.[17]

Historically, this has not been the case among the pioneers of quantum theory. In his paper entitled "'Mysticism' in quantum mechanics: the forgotten controversy" Juan Miguel Marin argues that in the period beginning with the birth of quantum theory to the middle of the twentieth century culminating in Schrodinger's speculations on mind and matter, there was intense controversy among leading physicists about whether quantum theory assigned a greater role to the mind in shaping the properties of matter than classical physics, which took for granted the objective existence of matter and its properties. Marin argues that in this controversy Planck and Einstein stood on the side of scientific materialism and objectivism and repudiated quantum theory for paving the way for such idealist conceptions. In contrast, Pauli and Schrodinger openly endorsed it as assigning a greater role to the mind in shaping the properties of matter. The culmination of such idealist positions can be seen in E.P. Wigner's consciousness interpretation of quantum theory, which saw quantum properties as acquiring definite values by virtue of their interaction with the mind of the observer.[18]

However, in the Anglo-American world there emerged a new kind of quantum mysticism which may be characterized as intuitionist rather than idealist driven by writers such as Fritjof Capra and Gary Zukav. This variant holds that consciousness or mind does not shape the reality observed by physicists, but that Eastern thinkers anticipated the discoveries reached by observation and experiment by scientists through introspection and intuition in mystic states of consciousness. This involved appealing to a sort of transcendental empiricism in place of the sensory empiricism of modern science. As a result Bohr has more recently come to be seen as embracing Eastern mystical doctrines by virtue of his interest in Eastern epistemological views. However, in the very passage that draws our attention to the epistemological parallels between quantum physics and Eastern philosophies, quoted at the beginning of this study, Bohr notes that his interest in Eastern thought is not by any means an endorsement of "mysticism foreign to the true spirit of science".[19]

It is significant that Bohr appreciates the importance of Eastern epistemological insights but does not endorse the quantum mystical and religious interpretations with which they have become connected. He does not claim, as quantum mystics do, that intuition allows us to apprehend the phenomena of the micro-world which physicists study, even if it leads to parallel epistemological viewpoints. Such a view is clearly reasonable. After all, even though great scientists, such as Newton and Kepler, were motivated by religious concerns to embrace certain epistemological and methodological positions, it does not stop us taking these positions earnestly. The same could be said of modern philosophers such as Descartes, Leibniz and Kant—also driven by religious impulses to develop and defend specific epistemological positions. We appreciate that what is important is that these thinkers defended their views by appeal to reason and evidence, and did not simply invoke religious authority or tradition.

Hence, we should be prepared to extend the same courtesy to Eastern thinkers and take their views seriously to the extent that they secure them by appeal to reason and argument, and not simply by appeal to the authority of revelation or religious tradition. In this context, it is also imperative to recognize that philosophical and religious ideas cannot be sharply separated in Eastern thought along the same lines as in the modern West—what is taken as a philosophical idea in the West can arise within a religious context in the East. This has been noted by the Japanese comparative philosopher Hajime Nakamura in his study *Parallel Developments: A Comparative History of Ideas*:

> In the West the two terms [religion and philosophy] have been fairly sharply distinguished from each other, while in Eastern traditions the dividing line is often difficult to discern. If we insist on being too strict in our definition, we fail to catch many common problems. It is possible that an idea or attitude held by a Western philosopher finds its counterpart not in an Eastern philosopher but in an Eastern religious thinker and vice versa. (Nakamura 1975: 3)

There is also support for Nakamura's observation in the generally accepted notion in Eastern cultures that the same tradition can be approached from a philosophical or religious perspective. Thus we have philosophical and religious Daoism, philosophical and religious Confucianism, and philosophical and religious Buddhism, often separated clearly in Eastern traditions. The religious variant tends to be associated with rites, rituals, beliefs, and practices often seen as less important by followers of the philosophical tradition. Even in traditions theologically more tied to orthodoxy of belief based on faith, such as Islam, there are important thinkers who distinguish between what can be termed prophetic and philosophical approaches to spirituality. Ibn al-Farabi, for instance, sees prophets as couching in metaphorical language that is appealing and comprehensible to the illiterate masses, the same knowledge that philosophers can discover through rational thought (Smart 2000: 165–166).[20] In all these cases the fact that a belief is associated with religion is not taken to taint it philosophically, provided it can be rationally defended on grounds independent of mere appeal to scriptural or theological authority.

But even those prepared to go along with Bohr's recommendation to open a dialogue with Eastern thinkers might suspect that such an engagement cannot illuminate the epistemological issues raised by quantum physics. There are a number of grounds for thinking that any engagement of this kind would turn out to be sterile. First, it is possible that the parallels noticed by Bohr between Eastern ideas and the epistemological implications of quantum theory are simply spurious. We may have 'discovered' such parallels only because we have selectively culled and combined passages from Eastern literature that bear some similarity to epistemological claims made by modern physicists. The coincidence of views we discern may have been constructed by foraging through a wide array of Indian and Chinese philosophical literature, taking material from such texts out of their contexts, and linking them together so as to fit our current views concerning the philosophical implications of quantum theory. In effect, we have constructed the parallels we claim to have discovered.[21]

The notion that the parallels Bohr noticed have been artificially constructed gains strength when we consider that there are other passages from the same Eastern texts that are not compatible with the findings of modern science. Hence the question arises as to why the selected passages from these texts should be given any epistemological or metaphysical priority over others that are ignored or rejected. This also provides further grounds for suspecting that the epistemological parallels with Eastern thought seen by Bohr are the outcome of sifting passages from ancient texts simply on the basis of their concurrence with the views of modern science or its implications. This has been stressed by a number of writers. For example, the sociologist Susantha Goonatilake argues:

> What Capra has attempted, in an ultimate sense, is to forage in the storehouse of Eastern thought and pick and choose elements that fit in with his own conceptual field and epistemological needs. He does not by this exercise discover an essentially new continuity in worn-out Eastern intellectual limbs, he only points to possible directions of absorbing the East from the perspective of the need for continuity of his own scientific tradition. (Goonatilake 1982: 271)

Moreover, such hindsight-based constructions do not appear too difficult to make. Given the vast number of available texts in Indian and Chinese philosophical traditions, and the long historical periods over which they were elaborated, and the proliferation of schools and sub-schools within them, it appears only too easy, by foraging for supportive textual material from this vast literature, to exhibit parallels to *any* philosophical position someone might care to adopt or invent—including that implied by quantum theory.[22]

The possibility that we have constructed the parallels we observe gains vigor when we take into account the profound gulf which separates the techniques, goals, and domains of inquiry that concern physicists and Eastern philosophers. Consider first their techniques. The physicists deploy highly elaborate, complex, and sophisticated instruments like ionization chambers, scintillation counters, and particle colliders. By contrast Daoist philosophers express aversion for all complicated machinery, disdain the method of controlled experiment, and recommend non-intervening communion with nature. They espouse a methodology that facilitates spontaneous development of natural processes without human interference. Buddhist philosophers are interested in neither experimentation upon nor communion with nature. Instead, they advise withdrawing from nature by adopting introspective techniques of meditation. Their path is alien to both the use of crafted technologies to study nature by the physicist and Daoist techniques for communion with nature. How can such different techniques, a skeptic could query, lead to similar epistemological concerns as Bohr implies?

The goals of the physicist and Eastern philosophers are also quite different. The aim of the physicist is to discover the universal laws that shape the particular phenomenon under observation. By contrast, the Daoist is not after general laws but only behavior within the specific contexts in which phenomena arise—contexts identified by living in close association with nature. The Buddhist philosopher is indifferent to both universal laws and context specific behavior that condition natural phenomena—they are both seen as illusory (*maya*), and the goal of the philosopher is to transcend the relative phenomenal world projected by the mind to apprehend the reality which sustains it. How could these divergent goals of physicists and Eastern thinkers, a skeptic could inquire, lead to similar epistemological concerns as Bohr implies?

Consider also the differences in the domains of inquiry that concern physicists and Eastern philosophers. The world of quantum physics is the world of atomic phenomena; the world of the Daoist is the natural wilderness found far from society in mountain and forest retreats untouched by human intervention; and the world of the Buddhist thinker is a psychological realm reached by introspective withdrawal from the empirical world of physical objects. How could such divergent realities, a doubter could easily wonder, lead to parallel epistemological viewpoints as Bohr implies?

But many New Age thinkers have not been daunted by such skeptical arguments. They contend that the epistemological parallels noticed by Bohr show that Daoist and Buddhist philosophers are apprehending the same world as the physicist, but perceiving it by different means.[23] The most prominent New Age thinker to have adopted this line of argument is the physicist and cultural critic Fritjof Capra. In his widely read and translated study, *The Tao of Physics: An Exploration of the Parallels Between Modern Physics and Eastern Mysticism*, Capra suggests that, notwithstanding ostensible appearances, physicists and Eastern thinkers are approaching a common reality, although scientists reach it through reason supported by empirical experience, and Eastern philosophers through, what he terms, the 'faculty of intuition'. Moreover, Capra considers the combination of sensory empiricism and reason in science to only apprehend this reality indirectly. By contrast Eastern thinkers directly encounter the same reality through intuition by suspending the operations of the discursive rational mind through the use of meditation and yoga techniques.[24]

Thus Capra sees the harmony between the views of Eastern thinkers and modern physicists to result from their adoption of different paths to arrive at knowledge of the same reality. He argues that the physicist "experiences the world through an extreme specialization of the rational mind", and the Eastern thinker "through an extreme specialization of the intuitive mind". Capra maintains that the fact that they reach the same conclusion despite "one starting from the inner realm, the other from the outer world" confirms the ancient wisdom that the ultimate reality seen without is the reality within us. Since Eastern thinkers and physicists are concerned with the same reality, so Capra argues, we cannot be surprised to find that the knowledge gained through nurturing intuitive awareness confirms and supports that derived from scientific exploration—intuition and empirical reason offer two different paths to one reality (Capra 1975: 323–324).

Capra's explanation for the parallels between the discoveries of physicists and Eastern thinkers is endorsed by the philosopher Renee Weber. In her book, *Dialogues with Scientists and Sages: The Search for Unity*, Weber goes even further than Capra. She explains that Eastern philosophers can actually perceive quantum phenomena directly by reaching an extraordinary state of consciousness that harmonizes their awareness with the subatomic matter that composes their bodies—a harmony that makes it possible for them to directly perceive, via intuition, the deep structures of physical reality that quantum physics investigates (Weber 1986: 12).

Both Capra and Weber assume that Eastern philosophers and modern physicists confront a common reality apprehended through different faculties—the introspective faculties of intuition and the faculties of the external senses that yield empirical experience. The assumption they make has come to be endorsed in much of New Age writing. However, this assumption raises more problems than it solves. Why was the picture of the atomic world achieved through intuitive apprehension not presented less ambiguously in the Eastern texts? Why are we left to extract this picture only with great difficulty from such texts, and furthermore only after, and in the light of, the empirical discoveries of modern science? Why did we have to wait until the twentieth century to arrive at a lucid picture of the atomic world if they were already to be found in Eastern texts? Surely this suggests that it is more sensible to think that Eastern thinkers and modern quantum physicists cannot be talking about the same reality.

The social anthropologist Sal Restivo notes that the same point has been made in a different way by the professor of biochemistry, science fiction writer, and popularizer of science, Isaac Asimov:

> His (Isaac Asimov's) reply to Capra is that if the Eastern sages know as much about the universe as physicists do, why not turn to a reading of the Taoist text to discover the unanswered questions of modern physics? Of course, Asimov argues, such a strategy would be futile. But then, if ancient texts can be 'properly understood' only after physicists have reached their conclusions, what scientific value do the ancient texts have? (Restivo 1982: 48, quoting Asimov 1979)

Moreover, Capra's and Weber's New Age arguments are also logically circular. They presume what has to be proven in order to develop their proof. They assume that Eastern philosophers and modern physicists are approaching the same reality through different faculties of intuitive and rational-empirical apprehension because of epistemological parallels in their views, but they also explain these parallels by assuming that they are the result of approaching a common reality by different means. Without offering us new grounds independent of the parallels, to show that these different approaches involve the same reality, their arguments are logically circular—they take the parallels as evidence for responses to the same reality, and presuppose a common reality to explain the parallels.

It is also questionable whether Eastern thinkers would concede that quantum physics is addressing the same reality they confront. Buddhist philosophers are likely to say that quantum physicists are actually concerned with the lower empirical reality, which they themselves transcend through meditation practice. They would question the notion that the physical techniques of scientists, however ingenious in design, and their theoretical understanding, which they see as an obstacle to intuitive knowledge, can ever lead beyond the empirical realm to the intuitive knowledge that is the Buddhist's ultimate illuminative goal. In this regard, most physicists would also concur with Buddhist philosophers, since they would agree that psychological and psychophysical techniques, such as meditation and yoga, cannot allow access to the world of atomic phenomena that they study. Hence, most physicists and Buddhist thinkers alike would agree that the goals and methods they adopt are so radically divergent that it is hardly credible to presume that they are approaching the same reality through different routes.[25]

What has been said of Buddhist philosophers can also be said of their Daoist counterparts. Like the Buddhist thinkers, the Daoists would hold that the knowledge sought by physicists is suspect. Daoists had come to reject the use of elaborate machines long before the industrial age at the time of the agricultural revolution in China when machinery was much simpler. They had objected to the use of mechanical devices for the intensive cultivation of nature because these interfered with natural processes. They had maintained that we cannot acquire knowledge of nature using mechanical contraptions to manipulate and distort natural contexts, but only by achieving a noninterventionist communion with nature in which natural processes reveal themselves. How much more would they reject the sophisticated instruments now deployed to study the quantum world? Indeed they are more likely to advice physicists to give up their experimental methodology based upon controlling natural processes, and study nature by communing with it without excessively interfering with its ways. Consequently, Daoists would reject the notion that physicists have come to apprehend the same reality that is also their concern—albeit by adopting different means.

Moreover, if Buddhist and Daoist philosophers did somehow manage to obtain access to the atomic world of quantum physics, we should expect them to give us a detailed account of this realm, which would add to, and possibly amplify, the discoveries of modern physics. Instead, what we find in their writings are at best only epistemological parallels to those implied by modern science. These parallels exist at the more abstract and general level of philosophical ideas rather than scientific descriptions. Hence, if these parallels have any significance, it is more likely that they are the result of being responses to similar properties seen in different objects in distinct domains of inquiry rather than responses to objects in a common domain apprehended by the different approaches of intuition and empirical reason.

This is a more reasonable and attractive alternative to the New Age notion that Eastern thinkers and modern physicists approach atomic events by different routes, because it does not require us to embrace the incredible notion that quantum physicists and Eastern philosophers are facing the same domain of objects. We only need to invoke the more reasonable assumption that they are led to similar epistemological views by virtue of confronting similar properties in objects within their different domains of inquiry. The parallels Bohr noticed between Eastern and complementarity epistemologies—which we will henceforth refer to as "the Bohr parallels"—can now be explained as responses to similar properties recognized by physicists in the natural world of atomic physics, by Buddhist philosophers in their intuitive experiences, and by Daoists in the world of wild nature.

The hypothesis that the Bohr parallels arise from the apprehension of similar properties in the quantum micro-world, the intuitive experience of the Buddhist philosopher, and the nature Daoists confront also circumvents many of the objections we confronted earlier against taking the Bohr parallels seriously. For now reservations concerning the existence of such parallels based on the differences in the techniques, the goals, and the domains of phenomena that concern scientists, Daoists, and Buddhist philosophers cannot be conclusive. All these three groups of practitioners could be led to comparable epistemological concerns because they are responding to similar properties in objects from quite different domains of inquiry and not properties of the same objects apprehended in a common domain of inquiry. The problem would be to identify the shared structure of the properties in the different kinds of objects confronted by scientists and Eastern thinkers that give rise to these epistemological parallels.

Explaining the Bohr parallels by assuming that scientists, Buddhist seers, and Daoist sages are confronting different kinds of objects with similar structure of properties would not only allow us to reject the notion that these parallels are hindsight-based and spurious constructions but also the explanation that they are responses to the same set of objects apprehended by different means. Such an account, however, still leaves open the question of identifying the structure of these properties that require us to adopt the epistemology of complementarity.

But before undertaking this task let us investigate another issue raised earlier—why the epistemology of complementarity has also come to be identified with postmodern social constructivism and antirealist views associated with it. Like its identification with New Age views this constitutes another obstacle against scientists and philosophers taking seriously the Bohr parallels and the wider implications of complementarity for science in general as envisaged by Bohr.

1.3 Complementarity and Social Constructivism

The notion of complementarity epistemology as implicitly endorsing a version of postmodern theory only developed significantly among natural scientists after the publication in 1994 of the study *Higher Superstition: The Academic Left and its Quarrels with Science* by the biologist Paul Gross and the mathematician Norman Levitt. These authors launched their book as an attack against a whole slew of environmental, feminist, and multicultural critics of science who they saw as adopting postmodern constructivist philosophies to articulate positions often inspired by the Copenhagen interpretation of quantum theory.[26] Gross and Levitt argue that these critics, whom they lump together as part of an "Academic Left", mainly grounded in the humanities and the social sciences, are motivated by a "higher superstition" inspired by misguided notions of the nature of science and its methodology (Gross and Levitt 1994: 3–4).[27]

What Gross and Levitt refer to as the Copenhagen interpretation is a position that combines Niels Bohr's complementarity perspective with Werner Heisenberg's famous uncertainty principle.[28] The Copenhagen view was developed by Bohr and Heisenberg together in order to rebut critics of quantum theory such as Albert Einstein. While Heisenberg had traced the uncertainty principle to the uncontrollable interactions of a measuring apparatus with an atomic particle when we set out to measure the properties of the particle, the Copenhagen view appealed to the uncertainty principle to defend the coherence of the complementarity framework against its critics. Despite decline in support for it, the Copenhagen viewpoint continues to remain the most widely accepted interpretation of quantum theory among practicing scientists. But Gross and Levitt express distaste for the Copenhagen viewpoint because they see it as motivating many postmodern social constructivist views of scientific knowledge. For example, they impute that Werner Heisenberg is embracing a form of 'mysticism' that inspires cultural constructivist views of science such as those espoused by the sociologist Stanley Aronowitz:

> [Aronowitz] insists on adverting only to the most mystical views of the matter (those of Heisenberg *qua* philosopher-oracle for instance) and ignores the particulars of the lively debate among physicists attempting to clarify what the predictive success of quantum mechanics really tells us about the physical universe. He naively echoes, for example, the view that the causal and deterministic view of things implicit in classical physics has been irrevocably banished. This is simply wrong. (Gross and Levitt 1994: 52)

The above quote makes it clear that Gross and Levitt dismiss Heisenberg as a 'philosopher-oracle', and recommend that we reject the Copenhagen interpretation. Instead they suggest we adopt the causal theory of quantum physics, first systematically articulated by the physicist David Bohm. In the process they actually end up adopting an even more controversial interpretation of quantum physics—one seen by most physicists as philosophically and scientifically less convincing (Gross and Levitt 1994: 262).[29]

Hostility to the epistemology of complementarity is taken even further by physicist Alan Sokal in his paper, *Transgressing the Boundaries: Toward a Transformative Hermeneutics of Quantum Gravity*. Sokal used this paper to perpetrate his (in)famous hoax on the cultural studies journal *Social Text* by arguing, tongue-in-cheek, the view that recent discoveries and trends in physics indicated a "foreshadowing of postmodernist epistemology". (Sokal 1996a: 220) However, shortly after publication of the paper in *Social Text*, he published another paper in a different cultural journal *Lingua Franca* revealing that his first paper had been intended to be nothing more than a hoax. In his new paper, entitled *Revelation: A Physicist Experiments With Cultural Studies*, he explains that he wrote the first paper not only to test the intellectual integrity of *Social Text* as a leading cultural studies journal but also postmodern cultural studies in general, by experimenting whether he would be able to publish an article "liberally salted with nonsense" by flattering editorial prejudices of cultural studies practitioners. He claims that his first paper was crafted solely to pander to the pride and prejudice of the cultural studies editors of *Social Text* who, although ignorant of science in general, were nevertheless flattered by his ostensible attempt to bring physics and postmodern theory closer. He achieved this goal not only by referring to quantum theory in his paper title, but also by using extensive quotes from Bohr and Heisenberg dealing with the uncertainty principle, quantum complementarity, and the Copenhagen interpretation. According to Sokal his second paper set out to expose the absurdity of the claims he made in his first:

> The fundamental silliness of my article [in the first paper] lies, however, not in its numerous solecisms but in the dubiousness of its central thesis and of the "reasoning" adduced to support it. Basically, I claim that quantum gravity—the still-speculative theory of space and time on scales of a millionth of a billionth of a billionth of a billionth of a centimeter—has profound *political* implications (which, of course, are "progressive"). In support of this improbable proposition ... I quote some controversial philosophical pronouncements of Heisenberg and Bohr, and assert (without argument) that quantum physics is profoundly consonant with "postmodernist epistemology." (Sokal 1996b: 62–64)[30]

The dismissive manner in which Sokal makes reference to the views of Bohr and Heisenberg in the above passage is revealing. Given the high esteem Bohr and Heisenberg enjoy within the science community, we might expect Sokal not to set out to repudiate their epistemological views as intellectually flawed, but only expose the silly interpretations postmodern writers had put upon them. If this were the case then Sokal's strategy would have turned out to be an ingenious subversion of postmodern attempts to misuse the philosophical views of Bohr and Heisenberg. Sokal would then be found to be on the side of Bohr and Heisenberg as he exposes the shallow relativism of postmodern humanists and social scientists, especially concerning the issue of how to correctly interpret the epistemological significance of quantum theory.

However, by dismissing the views of Bohr and Heisenberg as 'controversial', Sokal suggests that their philosophical views do lend support to the postmodern positions he is intent on sabotaging. He is not merely exposing postmodern misunderstanding of science—he is also opposing the Copenhagen position of Bohr and Heisenberg *per se*. Thus, Sokal is not merely engaging in the so-called 'science wars' between natural scientists and their postmodern critics; he is also a partisan in the contest within the natural science community between those supporting and others opposing the Copenhagen interpretation of quantum theory and its associated complementarity framework.

Sokal's anti-complementarity position is more evident in a letter, which he wrote in collaboration with another physicist Jean Bricmont, sent to the journal *Physics Today* in support of the historian of science Mara Beller's critique of Bohr's general epistemological standpoint. They write:

> She observes, correctly, that famous physicists such as Niels Bohr, Max Born, and Wolfgang Pauli engaged at times in dubious (to say the least) extrapolations of ideas from quantum physics to politics, psychology, philosophy, and religion; that these writings were sometimes treated by other physicists with excessive reverence, rather than being subjected to the critical analysis they deserved; and that the popular writings of these and other famous physicists—in which subtle conceptual and philosophical issues are often grossly oversimplified—have served as one source of inspiration (among many others) for postmodernist musings about science. (Sokal and Bricmont 1999: 15)

Sokal and Bricmont explain what they see as the "dubious extrapolations" of the complementarity perspective beyond physics to what they label as "the hubris that leads some (usually aging) physicists to enter into the "great minds" mode and imagine that whatever discoveries they made in physics must have deep consequences for philosophy or human affairs.". (Sokal and Bricmont 1999: 82) They vent their deep-rooted antipathy to the Copenhagen viewpoint by pointing out that "many physicists have for years blindly repeated Bohr's and Werner Heisenberg's views on the foundations of quantum mechanics, without having a clear idea of what they mean. We are pleased to note that the grip of the so-called Copenhagen orthodoxy is weakening and that physicists are beginning to consider alternative views on foundational questions with an open mind." (Sokal and Bricmont 1999: 82) Even more significantly, in a book published nearly a decade later entitled *Beyond the Hoax*, Sokal continues to maintain his repudiation of Niels Bohr's complementarity viewpoint.[31]

There need be little doubt that the deep animus against the complementarity viewpoint expressed by natural scientists, such as Gross, Levitt, Sokal and Bricmont, is driven by their perception that it facilitates postmodern efforts to subvert scientific rationality and objectivity. These scientists see themselves as defending the tradition of scientific rationality threatened not only by postmodern critiques but also by Bohr's efforts to recast the epistemological heritage of classical science through his complementarity framework. Their call gathered even greater momentum following the support of the Nobel Laureate in physics, Steven Weinberg. In his book, *Facing Up: Science and Its Cultural Adversaries*, Weinberg does not hesitate to blame Bohr and Heisenberg for fueling postmodernist appeal to quantum physics for support:

> It seems to me though that Sokal's hoax is most effective in the way that it draws cultural or philosophical or political conclusions from developments in physics and mathematics. Again and again Sokal jumps from correct science to absurd implications, without the benefit of intermediate reasoning. With a straight face, he leaps from Bohr's observation that in quantum mechanics "a complete elucidation of one and the same object may require diverse points of view which defy a unique description" to the conclusion that "postmodern science" refutes "the authoritarianism and elitism inherent in traditional science." (Weinberg 2001: 144–145)

Weinberg then goes on to commend the way "Sokal quotes some dreadful examples of Werner Heisenberg's philosophical wanderings" to illustrate how they have served as ammunition for postmodern excesses (Weinberg 2001: 147).

Although the epistemological science wars began as two quite distinct and different controversies—one as an internal conflict within the natural science community between those rejecting and those supporting Bohr's complementarity viewpoint and the other as an external conflict between natural scientists and postmodern interpreters of science—it is evident from the discussion above that many natural scientists have come to see these two debates one-sidedly as simply a confrontation between defendants of scientific rationality and postmodern critics who misunderstand and misappropriate science. This is reflected in the titles of the books they write for the wider public. Gross and Levitt's *Higher Superstition: The Academic Left and Its Quarrels with Science*, Sokal and Bricmont's *Fashionable Nonsense: Postmodern Intellectuals' Abuse of Science*, and Weinberg's *Facing Up: Science and Its Cultural Adversaries*, suggest that they have no issue with other natural scientists, but only with the camp of the Academic Left, intellectual postmodernists, or proponents of cultural studies. It takes a close scrutiny of their positions to discover that they are also against Bohr's complementarity framework deployed as a critique of the epistemology of classical science—a stance which pits them against the views of other scientists who adopt the Copenhagen position of Bohr and Heisenberg.

The lingering effect of this on Bohr's views is best exemplified by a passage from that excellent study *Who Rules in Science: An Opinionated Guide to the Wars* by philosopher James Robert Brown that serves as a bookend to the science wars. Brown attempts to combine scientific objectivity, values, and social influences in a way that does not subvert either scientific realism or scientific empiricism. Nevertheless, even in this attempt to forge a synthesis transcending the split between the two sides in the science wars, Bohr does not get a good press. Brown writes:

> As anyone who has struggled with Heisenberg and especially Bohr knows, these authors can be particularly obscure ... Wheeler and Zurek's *Quantum Theory and Measurement* is a standard reference work, collecting the most important articles on the foundational problems in quantum mechanics. One of the most famous is Bohr's reply to the Einstein, Podolsky, and Rosen paper. No one noticed that when reprinted in the Wheeler-Zurek anthology, Bohr's article had pages 148 and 149 transposed. After all, the original seemed a word salad, anyway. (Brown 2001: 94)

Nevertheless, scientists such as Gross, Levitt, Sokal, Bricmont and Weinberg, seem mainly concerned with portraying their principal targets as postmodern academic critics from the humanities and social sciences, rather than the great pioneers of science—Bohr and Heisenberg—even if they hold questionable some of the views of these founders of quantum theory. But the historian of science Mara Beller, standing outside the scientific community, has shown no such reticence. She charges that complementarity epistemology is incoherent, inconsistent, and logically untenable. She accuses Bohr of adopting it simply because it gave him the freedom to equivocate in presenting his views to different audiences, so as to project his authority and build solidarity among scientists on a fundamentally unstable position. This explains, says Beller, why Bohr has come to be understood so differently by many leading philosophers of science—a subjectivist by Karl Popper, an objectivist by Paul Feyerabend, a realist by Dugald Murdoch, and an antirealist by Jan Faye. Beller also thinks it explains why Bohr could claim to be a positivist in one context, and deny being one in another. Beller thinks that such dramatically different positions have been credited to Bohr because he developed his views in an *ad hoc* fashion designed only to make the different audiences he addressed experience a sense of kinship. It also explains, according to her, why such different groups as New Age thinkers, academic Left intellectuals, and radical anti-science feminists can embrace his ideas as a unifying factor to bind their otherwise diverse positions together.

In her paper *The Sokal Hoax: At Whom Are We Laughing?* Beller makes explicit the link she finds between Bohr's complementarity perspective and the excesses of postmodern theory:

> Astonishing statements, hardly distinguishable from those satirized by Sokal, abound in the writings of Bohr, Heisenberg, Pauli, Born, and Jordan. And they are not just casual, incidental remarks. Bohr intended his philosophy of complementarity to be an overarching epistemological principle—applicable to physics, biology, psychology and anthropology. He expected complementarity to be a substitute for the lost religion. He believed that complementarity should be taught to children in elementary schools. Pauli argued that "the most important task of our time" was the elaboration of a new quantum concept of reality that would unify science and religion. Born stated that quantum philosophy would help humanity cope with the political reality of the era after World War II. Heisenberg expressed the hope that the results of quantum physics "will exert their influence upon the wider fields of the world of ideas [just as] the changes at the end of the Renaissance transformed the cultural life of the succeeding epochs." … Sokal's hoax was ingeniously contrived. The gradual slide from the Bohr and Heisenberg quotes at the beginning of his article into postmodernist babble about the connection between science and politics is all too natural. (Beller 1998: 30–31)

It leads her to ask whether scientists who attack postmodernists should not also turn their sights on the pioneers of the Copenhagen view of quantum theory that they take to closely track the relativism and irrationalism of postmodern theory. Referring to the ridicule Gross and Levitt lay upon Aronowitz, simply because he assumes the end of determinism in science on the basis of the authority of some leading twentieth-century physicists, she asks:

> How can Aronowitz or other nonphysicists resist the authority of such past eminences, unless the physicists of our time publicly declare that the Copenhagen orthodoxy is no longer obligatory? Such a public declaration could have diminished greatly the explosive proliferation of the postmodernist academic nonsense so appalling to Sokal and Weinberg. The opponents of the postmodernist cultural studies of science conclude confidently from the Sokal affair that "the emperors … have no clothes." But who,

exactly, are all those naked emperors? At whom should we be laughing? (Beller 1998: 33–34)

Beller's question points directly to what can even be termed the 'internal science wars' within the natural science community that emerged with quantum theory. Beller challenges the scientific community to repudiate the views of the great physicists who adopted the Copenhagen interpretation—she thinks not doing so is providing ammunition for postmodern critics of science. But her argument is highly dubious. It endeavors to establish guilt by association—namely that those favoring the complementarity framework cannot be right simply because their views have been exploited by postmodern relativists. It is also the propensity to presume guilt by association, which motivates some scientists to pour rebuke and scorn upon eminent predecessors for embracing the Copenhagen standpoint. It encourages Gross and Levitt to charge Heisenberg with becoming "mystical" and turning "philosopher-oracle"; it makes Sokal and Bricmont invoke hubris to explain how leading physicists go astray by entering into a "great minds" frame as they age; and it prompts Weinberg to refer to "dreadful examples of Werner Heisenberg's philosophical wanderings."

Surely it is more reasonable to ask whether the pioneers of the quantum revolution who defended the epistemology of complementarity actually have something to teach us. Should we not give their views a more careful and considered hearing? Are we in danger of throwing the baby out with the bathwater when we reject their insights simply because they have been co-opted by postmodern thinkers?[32] Shouldn't we be investigating whether complementarity offers a new framework for scientific rationality, as Bohr maintained, without embracing the excesses of postmodern theory? Before launching *ad hominem* attacks against Bohr, as well as Heisenberg, shouldn't scientists, such as Gross, Levitt, Sokal, Bricmont and Weinberg, be asking: *At whom are we laughing*?

This question becomes even more pertinent when we consider the Bogdanoff Affair, which is sometimes characterized as the "reverse-Sokal" hoax. It involved the publication of a series of papers drawing upon quantum physics by the French twins, Igor and Grichka Bogdanoff, addressing theoretical issues about events that occurred at the Big Bang origin of the universe. These papers, going by titles such as "Topological field theory of the initial singularity of spacetime," in the journal *Classical and Quantum Gravity*, "Spacetime Metric and the KMS Condition at the Planck Scale," in *Annals of Physics*, "KMS space-time at the Planck scale," in *Nuovo Cimento*, and "The KMS state of spacetime at the Planck scale," in *Chinese Journal of Physics*, all involved purported applications of quantum theory to understand processes at the dawn of the universe. However, given the complexity of the topics they addressed including quantum groups and topological field theory, the review process floundered, and it later turned out that these papers had serious flaws that should have been detected. They were a hoax perpetrated on the physics community. Thus, if we are to reject complementarity simply because of the hoax perpetrated by Sokal on those who accepted this interpretation of quantum theory as endorsing postmodern views, shouldn't we also be rejecting quantum theory simply because of the hoax perpetrated by the Bogdanoffs on a number of leading physics journals? This would surely be throwing the baby out with the bathwater.[33]

The Sokal and Bogdanoff hoaxes also show why the debate between complementarity and its critics should not be cast in the form of a contest between natural scientists on one side and social scientists and humanists on the other. Indeed, complementarity may point to a position beyond the social constructivism of postmodernists and the objective realism of their critics. This is particularly the case because the notion of social constructivism is highly nebulous and its shifting meanings are difficult, if not impossible, to pin down. When we consider something to be socially constructed we imply that it does not exist in itself but by virtue of some social decision or choice. However, the nature of the decision is often not rigorously defined—it is variously interpreted as a choice of language, convention, belief, practice, and so on. Basically it is contrasted with the notion of essentialism—namely that something exists independent of human choice. Nevertheless, in his classic study *The Social Construction of What?* the philosopher Ian Hacking examines an extensive range of books and articles that carry titles having the form "The social construction of X" or "Constructing X". He finally concludes that saying something is socially constructed involves making at least two of the following claims:

1. X need not have existed, or need not be at all as it is. X, or X as it is at present, is *not* determined by the nature of things; it is not inevitable. (Hacking 1999: 6)
2. In the present state of affairs, X is taken for granted; X appears to be inevitable. (Hacking 1999: 12)

He uses this definition to deal with many natural and social categories such as facts, nature, quarks, child abuse, gender, and mental illness. Although

Hacking's work begins by directly connecting his study with the Sokal hoax, and the so-called 'science wars' between postmodernists and their critics, he does not directly address their connections to quantum physics.

However, Hacking's definition does cast light on why many have taken quantum theory as proposing that observed physical properties are socially constructed in contrast to the views taken for granted in classical physics. For example, the position and momentum of an atomic particle in classical physics is seen as existing independent of the measurement context, which merely reveals these values. But quantum physics, and the complementarity interpretation, requires us to assume that these properties do not pre-exist the measurement context and, depending on the measurement context, a particle can acquire a precise momentum or precise position, though not both at the same time. The position, say, did not exist at all prior to measurement—our choice of the measurement context brought it into existence. It is therefore socially constructed.

However, this argument is untenable. The fact that a property depends on an experimental arrangement based on a social decision does not make the observed property solely dependent on social choice. A property is a social construction only if it subsists by virtue of a social agreement, and ceases to exist the moment such agreement is withdrawn. But in the case of a measured quantum property, although social agreement decides what kind of measuring instrument would be deployed, social agreement does not create the property observed. The observed quantum property is the outcome of the interaction of the particle with the measuring instrument, and would not disappear after measurement even if we decide to revoke some social agreement. This is because the property observed is the outcome of an interaction with the physical apparatus, even though the choice of the apparatus to be deployed is based on a social decision.

Hacking himself is likely to endorse such a conclusion. He concedes that science itself has a history and can be, in that sense, seen as constructed, built and assembled. But he also wants to say that even if the idea of quarks may have a history, "quarks, the objects themselves, are not constructs, are not social, are not historical" (p. 30). He also deploys this "end of history" doctrine to defend the views of the physicist Steven Weinberg against that of the historian of physics, Norton Wise. Weinberg maintains that the second law of thermodynamics and Maxwell's equations, say, have been stripped off culture even if they arise within a culture, but Wise holds science and culture to be inseparably entangled. Hacking agrees with Weinberg: the thermodynamic law and Maxwell's equations "bear none of their history about them." (p. 87). Similarly quantum properties may be constructed, built, or assembled by our historical and social choice of the measurement contexts, but they may nevertheless, at the end of this history, exist independent of the social.

Attempts to argue that complementarity gives credence to postmodern social constructivism seem to succeed only if we do not distinguish the role of the physical context of the experimental arrangement in influencing observed properties, from the social context that selects the physical context. The measuring apparatus provides a physical context quite different from the social context involved in setting it up, and we cannot treat the *physical* context dependence of quantum properties as simply equivalent to *social* context dependence. Physical context dependence is a discovery of quantum physics that transcends classical science since classical properties pre-exist the physical measurement context, but quantum properties are shaped by the physical measurement context.

Postmodernists, who see dependence on social or linguistic contexts as similar to the physical context dependence discovered by quantum physics, ignore the novel implications of quantum theory. Indeed, since postmodern social constructivism applies to both classical and quantum physics it cannot serve to bring out the radical implications of complementarity epistemology, which is a response to the break quantum theory makes with the classical tradition.

In order to demonstrate this point, let us now consider three different attempts to articulate a postmodern interpretation of complementarity that, nevertheless, go beyond the realism and antirealism debates that characterized the science wars. The earliest of these is Arkady Plotnitsky's *Complementarity: Anti-Epistemology after Bohr and Derrida*, first published in 1994. It was followed by Karen Barad's *Meeting the Universe Halfway: Quantum Physics and the Entanglement of Matter and Meaning* (2007), and more recently by Makoto Katsumori's *Niels Bohr's Complementarity: Its Structure, History, and Intersections with Hermeneutics* (2011). All of these studies purport to articulate Bohr's complementarity perspective, and to claim that it transcends the realism and antirealism divide that was taken for granted in the science wars debates. More specifically, they perceive these debates to be undergirded by the assumption of a subject and object divide repudiated by Bohr's complementarity perspective.

According to Plotnitsky, Bohr maintains that "no independent physical reality or object can exist" (Plotnitsky 1994: 117), but he rejects the notion that Bohr's position can be treated as idealist or positivist or some other variant of antirealism. Plotnitsky argues that what Bohr calls reality cannot simply be reduced to observability, or treated as a construction of the subject, because there is what he designates as "material efficacity" that constraints observations, measurements, interpretations, and theories, which are not fully available to either observation or theoretical understanding.

He considers Bohr's concept of reality to be best understood as "alterity" in a Derridean sense. He writes:

> One might do well to abandon the term 'reality' altogether, provided that one takes precautions against positivist, idealist, phenomenologist, or transcendentalist interpretations that would reverse this concept without sufficiently displacing the metaphysical base grounding it. ... Rhetorically or strategically, alterity is a much better term. (Plotnitsky 1994: 108)

But Plotnitsky also stresses that alterity should not be interpreted in the Kantian sense of thing-in-itself, that is, it is not "absolute alterity". Bohr's position does not imply an absolutely other, but involves a complementary, i.e. a reciprocal, relation between subject and object, self and other, inside and outside.[34]

As a consequence Plotnitsky argues that Bohr, as with Derrida, rejects any "metaphysics of presence." (Plotnitsky 1994: 4) Such a metaphysics assumes a series of hierarchical binary oppositions such as self/other, inside/outside, presence/absence, speech/writing, and so on. Derrida uses the notion of *différance*—combining the notions of differing and deferring to point to the common root of all oppositional concepts so that one term of the binary opposition appears as the *différance* of the other, as the other different and deferred. (Plotnitsky 1994: 40–41) It leads Plotnitsky to characterize such paired terms as "complementary effects" arising from an operation he describes as "complementarization" (Plotnitsky 1994: 132).

According to Plotnitsky, Derrida reconceptualizes hierarchical binary opposites to become "heterogeneously interactive and interactively heterogeneous." (Plotnitsky 1994: 12) This is analogous, Plotnitsky argues, to Bohr's transformation of conceptual pairs such as subject/object and inside/outside. Although Plotnitsky's attempt to link complementarity to Derridean alterity appears somewhat forced, because he himself admits that there are differences between the two, since deconstruction in the Derridean sense is orientated toward "undecidability", whereas complementarity in Bohr is oriented toward "indeterminacy" (Plotnitsky 1994: 209), it is nevertheless clear that it builds on the interpenetration of the subject and object, or self and other. In a sense Plotnitsky's position can be seen as interpreting complementarity as transcending the realism and antirealism dichotomy presupposed by the contestants in the science wars.

Like Plotnitsky, the physicist and philosopher Karen Barad also interprets complementarity as pointing toward a transcendence of the subject–object dichotomy. She argues that it requires us to give up the notion of representationalism—namely that words and objects, or meaning and matter, are clearly separable. Instead she proposes a non-representationalist realism where the world is constituted by phenomena arising from what she terms the "intra-action", and not interaction, of objects and measuring agencies. The notion of intra-action is intended by Barad to point to the fact that the things or objects generally perceived as interacting are not prior to the phenomena observed, but arise through their intra-actions. She labels her position as 'agential realism' so that the apparatus that detect phenomena are seen as material-discursive, i.e. they produce both material beings and meaning at the same time. She writes:

> I propose "agential realism" as an epistemological-ontological-ethical framework that provides an understanding of the role of human *and* nonhuman, material *and* discursive, and natural *and* cultural factors in scientific and other socio-material practices, thereby moving such considerations beyond the well-worn debates that pit constructivism against realism, agency against structure, and idealism against materialism. Indeed, the new philosophical framework that I propose entails a rethinking of fundamental concepts that support such binary thinking. (Barad 2007: 26)

Barad supports her repudiation of binary thinking by arguing that Bohr maintained that quantum mechanics requires us to give up the distinction between subject and object so central to the classical world view. (Barad 2007: 359) Her agential realism is designed to reflect this transcendence of subject–object dualism. She writes:

> In agential realist account, scientific practices do not reveal what is already there; rather, what is "disclosed" is the effect of the intra-active engagements of our participation with/in and as part of the world's differential becoming. Which is *not* to say that humans are the condition of possibility for the existence of phenomena. Phenomena do not require cognizing

> minds for their existence; on the contrary, "minds" are themselves material phenomena that emerge through specific intra-actions. (Barad 2007: 361)

It is significant that although Plotnitsky and Barad both interpret complementarity in quantum physics as requiring us to repudiate binary thinking and the subject–object dichotomy, they nevertheless arrive at this conclusion by building their positions on the quite different notions of alterity and agential realism. Plotnitsky considers alterity to be oriented toward "undecidability"—in the Derridean sense—whereas Bohr's complementarity implies "indeterminacy." He recognizes that the two notions undecidability and indeterminacy are quite different, and that Bohr's indeterminacy is in some respects more radical than Derrida's undecidability. (Plotnitsky 1994: 2–5) He nevertheless perceives a resonance between deconstruction and quantum theory. By contrast Barad sees indeterminacy in quantum physics as a core basis for agential realism (Barad 2007: 359).

It is fruitful to contrast these two views with Makoto Katsumori's attempt to subvert binary thinking and transcend subject–object dualism by grounding complementarity on Bohr's metaphor of the spectator and actor perspectives. He writes:

> Bohr often characterized his idea of complementarity by the metaphorical dictum that 'we are both onlookers (or spectators) and actors in the great drama of existence.' In his view, while modern physical science has hitherto sought to see nature from the standpoint of a pure 'spectator,' the development of quantum theory has suggested that there can be no such purely detached standpoint, and that scientists themselves, as it were, unavoidably get involved in the drama of nature. More specifically, the observation of an atomic object carries with it an unavoidable and uncontrollable interaction with the measuring instrument—a circumstance which puts in question the conventional notion of independent objective reality. (Katsumori 2011: x)

Katsumori's position is much closer to Barad's than to Plotnitsky's. He admits this although her views are not considered in detail by him. (Katsumori 2011: 115) This is not surprising because Katsumori and Barad both take their positions directly off from Bohr, and the indeterminacy principle in quantum mechanics, whereas Plotnitsky grounds his position in Derridean undecidability and merely explores parallel notions in Bohr's complementarity. It leads Katsumori to criticize Plotnitsky for extending complementarity to include notions such as continuity and discontinuity, chance and necessity, and interiority and exteriority which will not be seen by Bohr as complementary because they are not mutually exclusive. (Katsumori 2011: 134) Nevertheless Katsumori also interprets Bohr's complementarity as subverting the subject–object dichotomy. He writes:

> Bohr's complementarity may perhaps be better characterised—without using the term anti-realism—as an undecidable suspension and alternation between the realist subject/object dichotomy and the non-realist disruption thereof. (Katsumori 2011: 71)

This leads Katsumori to reject, as Barad explicitly does and Plotnitsky tacitly, the implicit framing in the science wars debates that there can only be two possible philosophical orientations: realism involving belief in an independent external reality, and antirealism that treats everything as social constructs. (Katsumori 2011: 150) He argues that we have to reject such a dichotomizing of the material and the discursive or the natural and the cultural and emphasizes that others such as Barad and Latour have also argued along similar lines (Katsumori 2011: 157).

However, I would like to suggest that the attempt to treat complementarity as pointing toward the overcoming of the subject–object dichotomy is questionable. It fails to distinguish between properties that arise in the context of a physical environment, such as an apparatus used to measure quantum properties, and properties that arise by virtue of the cognitive framework we deploy to understand nature. The latter claim certainly suggests the interpenetration of subject and object in the way we cognize the world. This is really what is at the heart of what has come to be characterized as the "theory-ladenness" of observations in the so-called postmodern philosophies of science also linked to social constructionist views. However, this sort of interpenetration of the subjective and objective is as much applicable to the phenomena of classical physics as to quantum theory. It fails to capture the novelty of the epistemological feature of complementarity that requires us to transcend the classical tradition.

What is distinctive of quantum theory is that the properties we measure arise in the context of the measurement apparatus we deploy, unlike the case in classical physics where they pre-exist the context of measurement. The implication here is that the quantum property is shaped by the measuring context—it is the interaction with the physical environment that gives rise to it. This is the key discovery that Bohr's complementarity brings out—it shows why it involves a break with classical physics. Hence even if Plotnitsky, Barad, and Katsumori can interpret their positions as transcending the realism and antirealism dichotomy presupposed in the science wars, they nevertheless fail to capture the unique features of complementarity that make it applicable to the quantum world but not the classical one. In order to explore why this is the case we now turn to Bohr's attempts to extend complementarity into scientific disciplines beyond physics.

1.4 Psychological, Biological, and Anthropological Complementarity

This is quite a different project from the discovery of parallels to complementarity in Eastern thought, but it may also be seen as extending such parallels into other disciplines—in particular biology, psychology, and the social sciences. We also refer to such parallels as the Bohr parallels since, as we will later find, there is a connection between the parallels Bohr noted in Eastern thought and his project of extending parallels to other areas in knowledge. Bohr's efforts to take complementarity beyond physics can be found in numerous public talks and lectures given by him over a period spanning nearly four decades. Although many writers make references to Bohr's efforts in this direction, actual discussions of Bohr's views tend to be brief and superficial, sometimes even hostile, and mostly made only in passing when his epistemological ideas are mentioned within the context of his physics. This has made it difficult for readers to appreciate the significance Bohr attached to his wider epistemological project, or to perceive the coherence of his broad orientation.

However, Bohr himself did not underestimate the significance of complementarity beyond physics for knowledge in general. Bohr's interest in its wider epistemological lesson began very early after his Como lecture in 1927 in which he introduced his complementarity principle, which was published in the journal *Nature* seven months later. He writes:

> The very nature of the quantum theory thus forces us to regard the space-time co-ordination and the claim of causality, the union of which characterises the classical theories, as complementary but exclusive features of the description, symbolising the idealisation of observation and definition respectively. … Indeed, in the description of atomic phenomena, the quantum postulate presents us with the task of developing a 'complementarity' theory the con-

> sistency of which can be judged only by weighing the possibilities of description and observation. (Bohr 1928: 580)

What Bohr means by space-time co-ordination is the particle-like behavior of the atomic entity and what he refers to as the claim of causality is wave behavior. This becomes clear when he later refers to these as "complementary pictures of the phenomena" which "only together offer a natural generalisation of the classical modes of description" (See Pais p.315).

According to Bohr the description of our observations of atomic phenomena have to be made in the classical language of particles and waves despite the discovery of the dual aspect of wave and particle behavior by atomic entities in quantum physics. However, in classical physics picturing an entity as a wave excludes picturing it as a particle, so that a classical physicist would say that anything that exhibits one kind of behavior cannot exhibit the other. But in quantum physics it is otherwise. This dual aspect of atomic entities is most evident in the double-slit experiment. This involves using, say, a coherent laser light beam to illuminate a plate with two parallel slits through which it could pass. The light going through the slits is observed on a photographic screen on the other side. What we discover is an interference pattern on the screen of bright and dark bands due to the wave nature of light. But at the same time we also find that the light is absorbed by the screen at discrete points showing it to be made of individual particles. Indeed the interference pattern is constituted of the variations in the density of particles that hit the plate. The same result is obtained when the intensity of the beam is reduced so that only one atomic entity, on average, passes through the slits at one time. The discrete point absorption requires us to recognize the particle aspect of light, and the interference pattern shows the wave aspect, although it is not possible to combine these two pictures together at the same time. They are mutually exclusive visualizations but equally necessary for a comprehensive understanding of the double-slit experimental results.

However, there is another kind of complementarity that Bohr also raises in the Como lecture closely related to the wave-particle duality when he writes:

> "[A]ccording to the quantum theory a general reciprocal relation exists between the maximum sharpness of definition of the space-time and energy-momentum vectors associated with the individuals. This circumstance may

> be regarded as a simple symbolical expression for the complementary nature of the space-time description and the claims of causality."

This is the notion that the position and momentum of an atomic entity (or its duration in time and its energy) are complementary properties, in the sense that the precise determination of its position (or the duration) precludes the precise determination of its momentum (or energy). In contrast to the wave and particle duality, which are mutually exclusive from the point of visualization, and therefore excluded in classical physics, we have here a different situation. In this case properties, such as position and momentum, capable of visualization as mutually inclusive, and treated as such in classical physics, are no longer applicable at the same time to one and the same object. Central to such property complementarity is Bohr's view that choosing an experimental context, which makes it possible to precisely determine an entity's position, excludes any experiment that can precisely determine its momentum and also destroys what precise knowledge we may already have of it.

We will find that in order to appreciate Bohr's efforts to extend complementarity epistemology beyond physics it is imperative to separate these two notions of complementarity that he introduces. Both appeal to the notion of the "mutually exclusive but equally necessary", but the sense in which each is deployed is different. Wave and particle are mutually exclusive from a conceptual point of view in that to say something is a wave seems to exclude it as a particle, and vice versa, but, as we saw with the double-slit experiment, the two are equally necessary to apply to an atomic entity at the same time in order to account for its behavior. They can be seen as mutually exclusive double aspects of atomic behavior. Position and momentum are not mutually exclusive from a conceptual point of view, but are mutually exclusive in that they cannot be precisely measured at the same time. Nevertheless, both notions are equally necessary to account for atomic behavior. They can be seen as mutually exclusive properties of atomic objects. We will find that distinguishing aspect complementarity from property complementarity is central to understanding Bohr's efforts to extend complementarity beyond physics to which we now turn.

His interest in the wider applications of complementarity are evident in his 1958 volume *Atomic Physics and Human Knowledge* that brought together nearly three decades of his thinking about complementarity in areas as diverse as psychology, biology, and anthropology. In his Preface to the volume Bohr writes:

> The theme of the papers is the epistemological lesson which the modern development of atomic physics has given us and its relevance for the analysis and synthesis in many fields of human knowledge. (p. v)

He elaborates this further in his introduction to the volume where he writes:

> The main point of the lesson given us by the development of atomic physics is, as is well known, the recognition of a feature of wholeness in atomic processes, disclosed by the discovery of the quantum of action. The following articles present the essential aspects of the situation in quantum physics and, at the same time, stress the points of similarity it exhibits to our positions in other fields of knowledge beyond the scope of the mechanical conception of nature. We are not dealing here with more or less vague analogies, but with an investigation of the conditions for the proper use of our conceptual means of expression. Such considerations not only aim at making us familiar with the novel situation in physical science, but on account of the comparatively simple character of atomic problems be helpful in clarifying the conditions for objective description in wider fields. ... The gist of the argument is that for objective description and harmonious comprehension it is necessary in almost every field of knowledge to pay attention to the circumstances under which evidence is obtained. (pp. 1–2)

Although it appears to have been Bohr's intention to do an integrated work on the wider philosophical implications of complementarity—'The Book' as his close friends referred to it—he never did get around to it. However, we are fortunate that the 10th volume of *Niels Bohr-Collected Works* entitled *Complementarity Beyond Physics* brings together much of Bohr's efforts to show the relevance of complementarity epistemology beyond physics—especially for disciplines such as psychology, biology, and anthropology.[35]

Bohr's effort to apply complementarity to psychology followed closely on his Como lecture. We can find one of the most succinct accounts of the application of complementarity to psychology in his 1929 address in Copenhagen entitled "Atomic theory and the fundamental principles underlying the description of nature" at the *Congress of Scandinavian Scientists*:

> The fact that consciousness, as we know it, is inseparably connected with life ought to prepare us for finding that the very problem of the distinction

> between the living and the dead escapes comprehension in the ordinary sense of the word. That a physicist touches upon such questions may perhaps be excused on the ground that the new situation in physics has so forcibly reminded us of the old truth that *we are spectators as well as actors* in the great drama of existence. [My emphasis][36]

Bohr seems to argue that although psychologically understanding ourselves as spectators who are passive recipients of information coming in from the external world seems to exclude seeing ourselves as active agents constructing what we find in the world, descriptions of the standpoints of both the spectator and actor are required to fully understand psychological phenomena. Indeed, as the passage above suggests, in atomic physics at the quantum level, the result obtained when we measure a property is not only a function of what we perceive as spectators, but also a function of what measuring instrument we set up as actors.

However, Bohr's speeches, here and elsewhere, do not go beyond vague suggestions and illustrations as to how we could extend the complementarity of spectator and actor roles discovered in physics into psychology. Although he uses illustrations that appeal to the complementarity of subject and object, and free-will and determinism, they have not been either clear or cogent enough to convince others to take his extension of complementarity into psychology—what we shall term 'psychological complementarity'—seriously. But there is one area of psychological studies ignored by Bohr where the use of psychological complementarity of the actor and spectator perspectives can turn out to be illuminating—namely, gestalt psychology.

This is surprising because Bohr had discussed psychological problems with his second cousin Edgar Rubin and even helped him with experiments concerning visual perception for his doctoral thesis in 1915—only two years after Bohr had proposed his revolutionary quantum theory of the atom. (Bohr 1999: xxxix–xxx) Rubin had constructed the gestalt figure now known as 'Rubin's Vase' for his thesis on "Visually Experienced Figures". Since the figure can also be seen as two heads in profile, it demonstrated complementarity in perceptual experience—the perception of the vase and facial profiles are mutually exclusive, yet both are necessary in order to completely account for our experience of the gestalt figure (Bohr 1999: xlvi).

The theory of gestalts began with the discovery that what we perceive does not only depend on the external sensory stimulus, which can be taken as given objectively to us as passive spectators, but also the interpretation we choose as actors when reading such stimuli (which makes what we read depend on our free will as subjects).

Fig. 1.1 Duck-Rabbit Gestalt

For example, whether we perceive a duck or a rabbit in the above configuration of lines depends not only on the sensory stimuli we receive, but also the interpretive context we choose in responding to those stimuli. Choosing to interpret the figure as a duck we see X and Y as "beak" and "nape", but interpreting it as a rabbit we would identify X and Y as "ear" and "throat". Thus gestalts give flesh to Bohr's idea that psychology shows us to be both spectators and actors in the drama of existence. The perceived gestalt can be seen as presenting a double aspect—the configuration we confront as spectator and the interpretation we impose on it as actor. Although viewing the gestalt as externally imposed upon us as spectators and viewing it as mentally constructed by us as actors seem mutually exclusive points of view, both perspectives are necessary to fully comprehend the way we experience gestalts. They are complementary perspectives. The gestalt properties are understood as shaped by these two aspects as quantum properties are shaped by the wave and particle aspects.

Moreover, just as the quantum properties which arise from the interplay of the wave and particle aspects are mutually exclusive, so also are the gestalt properties which arise from the interplay of the configuration and the interpretation mutually exclusive. The duck properties exclude the rabbit properties and vice versa. Moreover to see the gestalt as the configuration or the interpretation does not give us a complete explanation of the rabbit phenomenon that we observe—it is only the dual aspect of configuration and interpretation that gives us such understanding.

It is noteworthy that the insights of gestalt psychology have come to transform the way philosophers of science see the nature of scientific observation. The discovery of gestalt psychologists brought an end to the notion that observation reports in science are simply passive responses to sensory stimuli. Our perceptions are gestalts that are shaped not only by sensory stimuli from an object but also theoretical interpretations we bring to bear upon them. Many philosophers of science now see all scientific observations as mediated by theoretical interpretations, i.e. as being theory-laden or theory-impregnated. As a result the observation of gestalts has important parallels with the observation of quantum properties. In particular, the gestalt properties we observe depend on the theoretical instrument we choose to deploy in order to read sensory stimuli, just as the quantum properties we observe depend on the measuring instrument we choose to deploy in order to make observations.

The importance of such parallels between gestalt and quantum properties, although strangely overlooked by Bohr, did not go unnoticed by some of the pioneers of gestalt psychology, including Max Wertheimer and Wolfgang Kohler. Addressing the issue of the logic of gestalt psychology in his address to the Kant Society in 1924 Wertheimer says:

> It has long seemed obvious—and is, in fact, the characteristic tone of European science—that "science" means breaking up complexes into their component elements. Isolate the elements, discover their laws, then reassemble them, and the problem is solved. All wholes are reduced to pieces and piecewise relations between pieces. The fundamental "formula" of Gestalt theory might be expressed in this way: There are wholes, the behavior of which is not determined by that of their individual elements, but where the part-processes are themselves determined by the intrinsic nature of the whole. It is the hope of Gestalt theory to determine the nature of such wholes. (Wertheimer 1924: 2)

He then goes on to say that what we need for psychological science is a mathematics that is not piecemeal. Admitting that finding such a mathematics would not be easy, he nevertheless expresses the hope that quantum theory would point the way in this direction.[37]

Kohler, another pioneer of Gestalt psychology, also connects this revolution in psychology with the quantum revolution in physics, and even claims that physicists such Max Planck, Max Born, and Bridgman endorse forging such a connection:

> In America, it may seem surprising that enthusiastic people such as the Gestalt psychologists were intensely interested in physics. ... organization is as obvious in some parts of physics as it is in psychology. Incidentally, others

> were no less interested in this "new reading" than we were. These other people were eminent physicists. Max Planck once told me that he expected our approach to clarify a difficult issue which had just arisen in quantum physics if not the concept of the quantum itself. Several years later Max Born, the great physicist who gave quantum mechanics its present form, made almost the same statement in one of his papers. And, only a few weeks ago, I read a paper in which Bridgman of Harvard interprets Heisenberg's famous principle in such terms that I am tempted to call him, Bridgman, a Gestalt physicist. (Kohler 1959)

These connections between gestalt psychology and quantum physics made by Wertheimer and Kohler are quite hazy, but nevertheless extremely suggestive, since they lend some weight to Bohr's attempts to extend complementarity from physics to psychology.[38]

Deploying the notion of gestalts to illuminate Bohr's notion of psychological complementarity of spectator and actor perspectives can also be used to explain why there are epistemological parallels to complementarity in Indian philosophical traditions. The epistemological views of many Indian philosophers were developed as responses to the plasticity of perceptual experience, which they achieved by entering altered states of consciousness induced by meditation techniques. Such techniques are designed to deconstruct the entrenched conceptual frameworks normally used to read sensory stimuli and organize everyday perceptual experience. It led many Indian philosophers to directly recognize the way our perceptual experiences are really conceptually constructed in response to external sensory stimuli—exactly what gestalt psychologists were to rediscover in the early twentieth century. It also led them to epistemological notions that parallel notions of complementarity in physics, as noticed by Bohr.

We will find this insight allows us to connect Bohr's extension of complementarity into psychology with parallels to complementarity in Buddhist and other traditions of Indian philosophy. We will also find that psychological complementarity has implications for the epistemology of science since it suggests that theories have the dual function of being instruments of observation as well as tools of inference. Indian thought had developed powerful techniques for dealing with theories as instruments of observation, but modern philosophy of science has shown how they can be potent tools of inference. As we will see later, complementarity in psychology makes it possible to bring these perspectives together in ways that have significance for philosophy of science in general.

Let us now turn to Bohr's attempt to extend complementarity to illuminate biological studies. His first effort in this direction can be traced to a talk entitled 'Light and Life' which he delivered in 1932 to an international congress of light therapists in Copenhagen. In the talk he addressed the question whether it would be possible to analyze living processes solely in terms of the laws of physics and chemistry. He argued that understanding living processes in an organism requires us to go beyond purely mechanistic terms and adopt a teleological approach to complement it. (Bohr 1958: 10) But Bohr's teleology does not involve assuming the existence of a separate material and vital substance—it is not a two-substance dualism. Instead he treats the mechanistic and teleological orientations as two equally necessary, but mutually exclusive, perspectives to fully understand biological processes. He says:

> A description of the internal function of an organism and its reaction to external stimuli requires the word *purposeful*, which is foreign to physics and chemistry … attitudes termed mechanistic and finalistic are not contradictory points of view, but rather exhibit a complementarity relation.[39]

In this passage, Bohr seems to suggest that a mechanical description in physico-chemical terms and a teleological description in terms of purpose at the level of the organism are both necessary to fully understand organic behavior. However, Bohr's extension of complementarity to biology in the above passage is as vaguely articulated as his attempt to extend complementarity into psychology. For example, he does not define clearly what he means by the term 'purposeful.' Does explanation involving appeal to purposes invoke conscious intentions and goals set by an organism? Or does such an explanation simply account for organic behavior in terms of how such behavior is adapted to promoting the survival of the organism in its environment? In short, does Bohr use 'purposeful' in the quote above to mean that the explanation must be in terms of the desires and goals that drives organic behavior or does he mean by 'purposeful' the functional role behavior plays in adapting the organism to the environment? Although Bohr is not explicit about the matter, it is reasonable to assume that Bohr is referring to the functional interpretation of biological processes since he never took two-substance dualism seriously. This would suggest that in order to understand all aspects of biological phenomena Bohr is recommending that we treat mechanical and functional approaches as complementary. We shall henceforth refer to this as Bohr's biological complementarity.

Nevertheless, mechanical and functional explanations of the structure and behavior of a biological system appear mutually exclusive. A mechanical explanation of a biological property or behavior explains it in terms of molecular properties and processes; i.e. genetic properties. By contrast a functional explanation of the property or behavior accounts for it in terms of the how it adapts the organism to its environment—i.e. the purpose served by the property or behavior in the overall successful functioning of the organism in its environment, i.e. the explanation is in terms of its environmental fit. Although the genetic and environmental explanations seem to be conceptually exclusive, both need to be deployed together in order to obtain a more complete understanding of the structural and behavioral properties of organisms. In short, the organism properties have to be explained by including the double aspect of gene and environment in a way analogous to explaining atomic properties by reference to both its particle and wave aspects.

Consider for example, the white color of the fur of polar bears. This can be explained in terms of the molecular structure of the inherited genetic code of polar bears. However, explaining it functionally in terms of the camouflage it gives the bear on its hunting expeditions in its environment adds to our biological knowledge. Although the molecular-genetic and functional-environmental explanations are mutually exclusive, they are both necessary. Given the complexity of biological processes, complete explanations in terms of molecular processes are impossible; we often need to complement such explanations with functional-environmental explanations to enrich our understanding. This suggests that the features of complementarity in the quantum domain that require us to use mutually exclusive, but equally necessary, descriptions are also applicable in the biological domain. Indeed, we will find that biological complementarity, precisely because it brings together genetic and environmental explanations, can greatly enhance our understanding of developmental, evolutionary, and ecological processes.

Consider the developmental processes through which a fertilized egg develops into a full-fledged mature organism through repeated cell multiplication. The division of the cells is influenced both by the shared genetic code as well as the environmental location in which it finds itself. Take, for example, a cell in the eye region of a growing frog embryo. Although it carries the same genetic code as a cell in the tail region it will not develop into a muscle cell. Hence, developmental biologists can explain its properties in two different ways—in terms of how these properties arise from physical and chemical interactions triggered by its genetic code and in terms of how these properties have been shaped by the environment in which the cell came to find itself. Although both these explanations appear conceptually exclusive, developmental biologists do not adopt one perspective to the exclusion of the other. Indeed the complex nature of the processes involved makes a full molecular explanation impossible, so that both perspectives have to be adopted to give a more complete explanation of the properties observed. Hence the molecular and environmental explanations have to be seen as complementary perspectives that allow us to give a more comprehensive account of the properties we wish to understand.

This has been emphasized by Scott Gilbert, in his popular text *Developmental Biology*, where he contrasts the reductionist and organicist views within the discipline. He writes:

> Both the reductionist and organicist approaches are materialist in that they do not invoke any extramaterial agent (entelechy; the soul; *Bildungstrieb*) as directing development. However, while reductionism claims that all complex entities (including proteins, cells, organisms, and ecosystems) can be completely explained by the properties of their component parts, organicism claims that complex wholes are inherently greater than the sum of their parts in the sense that the properties of each part are dependent on the context of that part within the whole in which it operates. Thus when we try to explain how the whole system behaves we cannot get away with speaking of the parts. (Gilbert 2010: 618)

Similarly, complementarity perspectives can also illuminate general features in observed patterns of organic evolution. The pattern of organic evolution has been the focus of intense debates since the nineteenth century. Some biologists think evolutionary changes take place gradually in small incremental steps (gradualism), but others see them proceeding in large disjunctive leaps (punctuationism). Gradualism and punctuationism give radically different pictures of the pattern of organic evolution. However, gradualism and punctuationism can be seen as descriptions of the same processes seen from two different perspectives. Viewed from a molecular perspective, in terms of the evolution of the genotype (genes) organic evolution is a gradual process. This is because the changes in the genetic structure of organisms over time generally involve small mutations, since large genetic mutations are likely to preclude the organism from either developing or reproducing itself. By contrast, viewed from the perspective of properties and processes of the grown organism or phenotype, organic evolution exhibits a punctuated pattern. This is because there are occasions when even small changes in the genetic code can become amplified into radical changes in the grown phenotype. The phenotype survives only if these changes are adaptive to the environment within which it finds itself. Moreover, even the genotype can propagate itself only if the phenotype it produces is adapted to its environment. Consequently, the phenotypic properties can be explained from two different points of view—in terms of the gradual changes in its genetic code and as a punctuated structure adapted to the environment. Indeed both viewpoints are necessary to completely understand the properties of the phenotype.[40] The genetic and environmental viewpoints are complementary perspectives on the phenotype that appear mutually exclusive since a molecular explanation of the phenotype properties seems to render unnecessary an environmental explanation, and vice versa.

However, we cannot treat cumulative and punctuated accounts of evolution as exhibiting complementarity—they are complementing perspectives since they are mutually inclusive patterns, one at the level of the genotype and the other at the level of the phenotype. But these complementing perspectives can only be explained by combining the double aspect of genetic and environmental explanations.

Biological complementarity of molecular and environmental explanations can also be extended beyond the development and evolution of individual organisms and species to giving an account of the ecological properties of natural systems in the biosphere. Consider the properties of an ecosystem, say the Amazonian rainforest. These properties can be explained in terms of how they developed through the interaction of its molecular parts over long historical time or, functionally, in terms of the role of the Amazonian rainforest as an environment maintaining conditions suitable for life on earth. The functional approach is evident in James Lovelock's *Gaia hypothesis* where he adopts what he terms a 'geophysiological' perspective on nature—one which sees any ecosystem in terms of its function in maintaining the conditions for life within the Earth's biosphere as a whole.[41] Hence, the properties of the Amazonian rain forest may be explained, in physical and chemical terms, by reference to the geological processes that brought them about or, in functional terms, by invoking its geophysiological role of serving as a 'lung' to maintain the life-supporting composition of gases in the atmosphere.[42]

Hence, treating as complementary the molecular and functional orientations can illuminate and enrich our understanding of processes in developmental, evolutionary, and environmental biology. We will find that such an approach can vindicate Bohr's view that complementarity can be extended into the biological sciences to enhance our understanding of living systems.[43]

Let us now turn to Bohr's attempt to extend the framework of complementarity into the social sciences. Bohr did not develop his views very far in this direction. One of Bohr's earliest attempts in this direction occurs in an address to the *International Congress of Anthropology and Ethnology* in 1938. In his talk, entitled 'Natural Philosophy and Human Cultures,' he presents his ideas on how the role of instinct and reason shape human behavior. He said:

> [T]he amazing capacity of so-called primitive people to orientate themselves in forests or deserts, which, though apparently lost in more civilized societies, may on occasion be revived in any of us, might justify the conclusion that such feats are only possible when no recourse is taken to conceptual thinking. (Bohr 1958: 28)

Thoughts like these led Bohr to see human behavior as explained by the complementary standpoints of nature (instinct) and culture (conceptual thinking). To see behavior as due to nature is to view it as independent of culture; to see it as the outcome of nurture is to see it as culturally controlled. Yet in explaining many human traits it is impossible to sharply separate the part that is the result of nature from the part that is the outcome of culture, since often both instinct and conceptual thinking play a role in shaping human behavior.

This realization led Bohr to maintain that human behavior can be viewed from two complementary standpoints—as shaped by instinctive biological inheritance or as shaped by cultural conceptual influence. According to Bohr:

> [I]n characterizing different nations and even different families within a nation, we may to a large extent consider biological traits and [traits rooted in] spiritual traditions as independent of each other, and it would even be tempting to reserve by definition the adjective "human" for those characters which are not directly bound to bodily inheritance. (Bohr 1958: 29)

The philosopher, Henry Folse, also argues that Bohr's distinction between the anthropological and the spiritual in the above passage is intended to characterize the nature versus nurture distinction. (Folse 1985: 174) Hence, it is reasonable to assume that Bohr maintains that any human behavior can be explained in two different ways—either as grounded in what has been given by nature or as produced by nurture. Although these are mutually exclusive explanations, both are necessary to fully understand human behavior—they are complementary explanations of such behavior. We will refer to it as anthropological complementarity to reflect both the biological and cultural dimensions it brings together as two aspects explaining human behavior.

The issue of whether we should follow nature or culture was also central to classical Chinese thought, especially in the debates between Confucian and Daoist thinkers. The crucial question that confronted these thinkers was how we should go about living our lives. Should we adopt the principle of yielding to our biologically inherited human nature or should we adopt the principle of culturally cultivating human nature? Confucians recommended the more assertive approach of culturally cultivating human nature—what was termed the masculine or *yang* orientation in Chinese culture—and Daoists valorized yielding to our natural human nature—what they termed the feminine or *yin*, orientation.

In contrast to the Confucians and Daoists, we have seen that Bohr advocates understanding human behavior through the complementarity of the culturally-shaped assertive masculine (*yang*) and the biologically-inherited yielding feminine (*yin*) perspectives. Hence, it is not surprising to find Bohr connecting his complementarity perspective with the Chinese complementarity of yin (feminine) and yang (masculine) principles. Indeed when he received his nation's highest honor in 1947—the *Order of the Elephant*—he had to design a crest and motto for his shield. His crest was the Chinese *taijitu* symbol for *yin–yang*, and his inscribed motto in Latin read *Contrari Sunt Complementa* (*Opposites Are Complements*).[44] His choice is significant because it shows how closely he connects his notion of complementarity to the yin–yang complementarity of Chinese thought.

Bohr's choice may not be a direct afterthought following his discovery of complementarity—it is likely that the notion of complementarity may even have been influenced by his early acquaintance with Daoism. According to Robert Allinson, there is evidence that Bohr learnt of the

Chinese philosophy of Lao Tzu at an early age from new evidence in the form of a letter written by Bohr himself. Allinson adds:

> His metaphor of complementarity is the expression of the basic principle of *Yin-Yang* philosophy. If a metaphor taken from Chinese philosophy may be of great explanatory power in the world of physics, may it not be all the more possible that the same metaphor may possess great explanatory power in the world of philosophy? This might represent a case in which a regulative principle of physics can be utilized as a regulative principle of philosophy. (Allinson 1998: 507)

Moreover, just as psychological complementarity can be linked to Buddhist epistemological complementarity, we will find later that anthropological complementarity in Chinese thought can be linked to biological complementarity of the molecular and functional perspectives. Indeed, the functional perspective suggests that biological entities are nurtured by the context in which they develop as much as the molecular processes within them. This suggests that when we deal with biological nature we can decide either to cultivate the context in which things grow or let things grow spontaneously within their natural context.

This became an issue of intense controversy in the Chinese world at the dawn of the Axial Age. Confucians advocated the economic exploitation of nature through intensive agricultural cultivation to enhance human welfare. Their assertive (*yang*) perspective was resisted by Daoists, who recommended the feminine (*yin*) principle of not disturbing natural ecological processes through human interference. But economic activities and ecological natural processes both produce goods and services necessary to sustain human life and well-being. As a result, we will find that the anthropological complementarity of the *yin–yang* (yielding–assertive) approach that Bohr defends can be used to illuminate concerns about how to engage in economic activity without violating natural ecological integrities—an issue that we examine in greater detail in the fourth chapter.

There is a common pattern that informs Bohr's understanding of complementarity in the quantum, psychological, biological, and anthropological domains. It ties in with the discovery in atomic physics that we have to do away with deterministic causal explanations and replace them with explanations that are mutually exclusive but equally necessary. In fact he argues that "the viewpoint of complementarity forms indeed a consistent generalization of the ideal of causality." (Bohr 1958: 27) As such it has implications for our understanding in areas of knowledge far beyond physics. Since the notion of deterministic causal explanations was closely linked to the mechanical world-view of classical physics Bohr was motivated by his belief that the complementarity principle had a significance for science as far reaching as the mechanical vision of the seventeenth-century Scientific Revolution. In the introduction to his book *Atomic Physics and Human Knowledge* Bohr writes:

> "The main point of the lesson given us by the development of atomic physics is, as is well known, the recognition of a feature of wholeness in atomic processes, disclosed by the discovery of the quantum of action. The following articles present the essential aspects of the situation in quantum physics and, at the same time, stress *the points of similarity it exhibits to our position in other fields of knowledge beyond the scope of the mechanical conception of nature*. We are not dealing here with more or less vague analogies but with an investigation of the conditions for the proper use of our conceptual means of expression. Such considerations not only aim at making us familiar with the novel situation in physical science, but might on account of the comparatively simple character of atomic problems be helpful in clarifying the conditions for objective description in wider fields." (Bohr 1958: 1–2) [My emphasis]

The passage above emphasizes three significant points. First the feature of wholeness recognized in atomic physics has parallels in other areas of knowledge. Second this is not a matter of vague analogies but directly linked to the logic of the conceptual means we have on hand. Thirdly the approach for addressing problems in the simple systems of atomic physics can be extended to wider fields of knowledge. Let us now examine how this connects quantum complementarity with psychological, biological, and anthropological complementarities.

First, in all of these areas of knowledge we have objects whose properties need accounting analogous to the properties observed in a quantum entity—psychological gestalts, biological phenotypes, and anthropological traits. Second, these properties are influenced by an entity or structure that we can refer to as a template that interacts with a context in which it is set—a configuration in a theoretical frame of interpretation, a genotype in an environment, and an inherited human nature within an embedding culture. The properties we observe in gestalts, phenotypes, and human beings are the outcome of the interaction of the templates—configuration, genotype, and inherited human nature—with the embedding contexts—theoretical frame, environment, and culture. Consequently explaining these observed properties solely in terms of the template or its context is incomplete—both the template and context underdetermine the properties observed. Yet, explanations in terms of the template and context are also mutually exclusive in that a complete explanation in terms of one precludes the need for an explanation in terms of the other. There is a close affinity here to the wave and particle explanation of quantum properties. It is this insight that leads Bohr to argue that "the comparatively simple character of atomic problems might be helpful in clarifying the conditions for objective description in wider fields". I realize that I may have read more into Bohr than he intended but I hope to show that such a reading is viable and justifies his claim that the discovery of complementarity in atomic physics has lessons to offer us for dealing with knowledge in disciplines beyond physics.

1.5 Complementarity and Grown Properties

Up to this point, we have only given in a sketchy fashion how we intend to develop Bohr's hints to widen the reach of complementarity beyond physics. However, we have seen that the Bohr parallels are also best explained not by assuming that physicists and Eastern thinkers are approaching the same phenomena through different paths but that they are responding to similar properties in different domains of phenomena. We have also pointed out, albeit in a sketchy fashion, directions for articulating Bohr's biological, psychological, and anthropological complementarities, as well as how these may link to notions of complementarity in Eastern thought. However, we have not addressed the question of what these properties are. Indeed the preceding section that sees these properties as arising through the interaction of a template with a context already hints at a direction for identifying these properties. I would like to propose the *hypothesis* that the properties we are concerned with are grown properties in nature, such as are found in the biological realm where things grow in particular contexts, and that treating quantum properties as grown properties can explain many of the features of complementarity that now seem puzzling.

Such a hypothesis may seem *prima facie* patently flawed, since even classical physics acknowledges grown properties in nature. However, what we want to suggest is that classical physics misidentified the structure of these properties. In order to appreciate why this came about we have to look at properties in simple mechanical systems, such as the mechanical clock that defined the classical vision of the universe. Indeed the clockwork mechanism constituted one of the most sophisticated systems for seventeenth-century scientists and philosophers and inspired the clockwork metaphor of the universe. Such a clock consists of a large number of component parts that are placed in well-defined relations to each other. Consequently the parts that make up the system have two different types of properties. First, they have properties that they had possessed prior to their becoming embedded in the context of the mechanical system. For a weight-driven pendulum clock, the kind of mechanism that inspired the clockwork metaphor, these properties would be the mass, shape, and size of the pendulum, the gear wheels, the pins that release the striking train, the hammer, the fly, the rack, the cam, the locking lever, the anchor escapement, and so on (Bruton 1989: 71). These are properties that would also be retained by the parts after they are removed from the system. We can describe mass, shape, and size as intrinsic properties of the parts because they originate and subsist in the parts independent of their location in the system.

Second, there are properties the parts acquire by virtue of their links with each other within the context of the simple mechanical system. It is precisely the pattern of relations that these parts have to one another that makes the clock function as an organized system that can keep time. Such properties, moreover, are lost by the parts when they are removed from their place in the system. We can describe these properties as relational properties of the parts because they originate and subsist in the parts dependent upon their position in the system.

When we assemble and disassemble a simple clockwork mechanism, or any other simple mechanism for that matter, we will find that the parts of the system have only these two kinds of properties, and that the effective functioning of the system is not due to the intrinsic and relational properties of its parts separated from each other, but to the way they come together and interact within the system. The system functions as a result of the mutual interplay of the intrinsic and relational properties of its parts. Moreover, there is no third kind of property in a simple clockwork system over and above the intrinsic and relational properties of its parts.

Most scientists of the seventeenth century who developed the new mechanical philosophy also considered the clock to be an ideal metaphor for all processes in the universe.[45] Consequently one of the characteristic epistemological and methodological orientations of the world conception that inspired early modern science was to see complex living things as mechanical systems. Descartes was one of the pioneers and leading figures in consolidating this approach through his endeavor to recast our image of living things in the clockwork metaphor. He writes in his *Principles of Philosophy* first published in 1644:

> And, to this end, things made by human skill helped me not a little: for I know of no distinction between these things and natural bodies, except that the operations of things made by skill are, for the most part, performed by apparatus large enough to be easily perceived by the senses: for this is necessary so that they can be made by men. On the other hand, however, natural effects almost always depend on some devices so minute that they escape all senses. And there are absolutely no judgments {or rules} in Mechanics which do not also pertain to Physics, of which Mechanics is a part or type: and it is as natural for a clock, composed of wheels of a certain kind, to indicate the hours, as for a tree, grown from a certain kind of seed, to produce the corresponding fruit. Accordingly, just as when those who are accustomed to considering automata know the use of some machine and see some of its parts, they easily conjecture from this how the other parts which they do not see are made: so, from the perceptible effects and parts of natural bodies, I have attempted to investigate the nature of their causes and of their imperceptible parts. (Descartes 1982: 285–286)[46]

However, by assuming that a running clockwork and a growing tree have no differences except that of scale, as Descartes does in the quote above, the clockwork vision is led to overlook the distinctive structure of properties that grow in biological systems. Such properties are neither intrinsic nor relational but constitute a third class of properties that may appear intrinsic or relational depending on the context in which we view them. Grown properties can be described as being relational or intrinsic when seen from different contexts, although both descriptions are necessary for fully understanding the behavior of such properties.

To appreciate the distinctive structure of properties that grow consider the height of a tree. Is this an intrinsic property of the tree or a relational property? It may appear to be an intrinsic property because it is not lost by the tree even when we remove it from its environment. However, the height of the tree is also dependent on the environment in which it has grown. If it had grown in an environment amply supplied with fertile soil, water, and sunlight it is likely to be tall, but if one or more of these factors had been in short supply it might have been shorter. Hence the tree's height also appears to be a relational property that originates in dependence upon the environment of its growth. Moreover, intrinsic and relational properties also appear mutually exclusive. By virtue of how it has been acquired in dependence on its environment a grown property is not intrinsic; and by virtue of how it is retained outside its originating environment it is not relational. Thus the grown property of the tree exhibits intrinsic and relational features.

Much confusion can be avoided if only we treat the height of the tree as a property distinct and different from both intrinsic and relational properties, but exhibiting features of both. This follows from the fact that it originates in dependence on the environment, as an intrinsic property does not, and subsists independent of the environment, as a relational property does not. Such properties do not exist in the simple mechanical systems studied by classical science in which things don't grow. However, grown properties are ubiquitous in the complex systems of the biological world.

In this study we will find that the hypothesis that quantum properties are grown properties can explain the various elements of the Copenhagen interpretation, and its complementarity perspective, in an elegant fashion. It also shows why the complementarity perspective developed to deal with quantum properties can be extended to the biological realm in general. Furthermore, it shows that Bohr's attempts to extend the framework of complementarity to illuminate phenomena in psychology and the social sciences are not misguided either. These extensions can be understood as taking into account our psychological and social responses to grown properties in nature. Even Buddhist and Daoist epistemologies can be ultimately traced to the responses of Eastern thinkers to grown properties in the natural environment. The rest of this study attempts to defend these claims, and thereby Bohr's view that complementarity has wider implications for science and philosophy beyond physics.[47]

Moreover, the extension of the complementarity principle beyond physics also has epistemological implications for science in general, not envisaged by Bohr. In the subsequent chapters we not only find that biological complementarity can be extended to illuminate the self-regulating ecosystems perspective as encapsulated in the Gaia hypothesis but also that psychological complementarity can contribute to a wider conception of the philosophy of science that acknowledges the interdependence of conception and perception, which can be enriched by incorporating insights from Indian epistemology, and that anthropological complementarity has implications for making economic theory sensitive to ecological constraints, and show why we can learn from Daoist epistemology to make this reorientation.

Notes

1. This principle was articulated and introduced by Niels Bohr in 1927. He summarized it as follows:

   > [H]owever far the [quantum physical] phenomena transcend the scope of classical physical explanation, the account of all evidence must be expressed in classical terms. The argument is simply that by the word "experiment" we refer to a situation where we can tell others what we have done and what we have learned and that, therefore, the account of the experimental arrangements and of the results of the observations must be expressed in unambiguous language with suitable application of the terminology of classical physics.
   >
   > This crucial point ... implies the impossibility of any sharp separation between the behaviour of atomic objects and the interaction with the measuring instruments which serve to define the conditions under which the phenomena appear .... Consequently, evidence obtained under different experimental conditions cannot be comprehended within a single picture, but must be regarded as complementary in the sense that only the totality of the phenomena exhausts the possible information about the objects. (Bohr 1949: 209–210)

2. This may be interpreted as a creation-at-measurement view. Such a view has been proposed by Clifford Hooker (1972).
3. *Aristotle's Metaphysics*, 4.4 in Ross (2006).
4. For a recent discussion of these issues see Beckwith (2015). He notes that the tetralemma is a figure that features even in classical Greek discussions of logic by Plato and Aristotle in their critique of the skeptics, but is largely rejected by them. See pp. 203–205. However, Kuzminski (2008) argues that the Greek tetralemma may have been influenced by the Buddhist *catuskoti*. It is also important to keep in mind the Buddhist *catuskoti* is a denial of the four positions expressed in the tetralemma. See also Jayatilleke (1967).
5. David Loy (1988) suggests that "the action of no action" paradox is the most fundamental and that the other paradoxes—"the morality of no morality" and "the knowledge of no knowledge" are more specific manifestations of the general pattern it establishes (Loy 1988: 97).
6. Many other physicists involved in the development of quantum theory also recognized that its philosophical implications seem to have parallels with certain traditions in Eastern thought. See Marin (2009). These views have, in recent years, been popularized by writers such as Fritjof Capra (1975) and Gary Zukav (1984) who have twisted it in a New Age spiritualist direction that Bohr would not have approved of. In a more recent study, the

physicist Arthur Zajonc (2004) brings together scientists, scholars, and religious thinkers to exchange comparative views on the philosophical implications of quantum physics and Buddhist philosophy—an event that shows the continuing interest that links these otherwise disparate traditions together.

7. See Birkhoff and Von Neumann (1936). Also Putnam (1968) and Dummett (1976). Reichenbach (1944) had argued earlier along similar lines. Sarukkai (2005) maintains that the Indian tradition had generally treated logic not as a discipline prior to empirical science but as a part of it. See especially Chap. 5 "Science in Logic: The Indian Way?" pp. 157–208.
8. It is worth noting here that Bohr was to replace the term 'vitalistic' with the term purposeful after the discovery of the double helix by Watson and Crick in 1953. Pais explains this change as follows:

   > Bohr's position had to change, of course, after the discovery in 1953 by Francis Crick and Jim Watson of the structure of DNA and of the physico-chemical processes of biological replication initiated by that specific molecule. All references to vitalism now vanishes from Bohr's writings on biology—but complementarity persists. (Pais 1991: 443)

9. In the inaugural volume of 1963 on Bohr, produced after his demise in 1962, the physicists Leon Rosenfeld and John Wheeler both emphasize this revolutionary dimension complementarity held for Bohr. According to Rosenfeld "complementarity is not a philosophical superstructure invented by Bohr to be placed as a decoration on top of the quantal formalism, it is the bedrock of the quantal description." (Rosenfeld 1996: 284–285) Similarly Wheeler described the status of the principle of complementarity for Bohr as follows: "Bohr's principle of complementarity is the most revolutionary scientific concept of this century and the heart of his fifty-year search for the full significance of the quantum idea" (Wheeler 1963: 30).
10. Thus Schrodinger maintained that "Bohr wants us to complement away all difficulties." Einstein intimated to Schrodinger that "The Heisenberg-Bohr soothing philosophy—or religion?—is so finely chiseled that it provides a soft pillow for believers ... This religion does damned little for me." (Both quoted in Pais 1991: 425).
11. Quantization refers to a procedure for using classical field theory to construct quantum theory by using the concepts of classical physics. This generalizes the procedure that Max Planck used when he quantized the electromagnetic field into packets that are field quanta or photons. Planck's proposal that action is quantized required revising our understanding of the use of classical concepts—they could not all be applied at the same time. Only in those situations where the quantization of action is negligible can we deploy them in the classical manner.

12. Hence, complementarity has also come to be connected with an earlier debate that began in the nineteenth century when a whole slew of Romantic critics inspired by social, humanist, and religious interests came to pit themselves against science and its epistemology. The Romantics appealed to philosophy, history, literature, and Eastern thought as approaches to transcend what they perceived to be the suffocating vision of mechanical science. Between the clash of scientists with the Romantics in the nineteenth century, and contemporary neo-Romantics, we have the divide between the humanists and the scientists bemoaned so brilliantly by C.P. Snow in his influential Rede Lecture in 1959. See Snow (1960).
13. A vast body of literature has come to be associated with these contestations—Kuhn (1970), Gross and Levitt (1994), Sokal and Bricmont (1998), Hacking (1999), and Brown (2001). Some leading critics of scientific realism have come to recant their views because they saw their position as lending intellectual ammunition to reactionary interests. For example, Bruno Latour writes: "[D]angerous extremists are using the very same argument of social construction to destroy hard-won evidence that could save our lives. Was I wrong to participate in the invention of this field known as science studies? Is it enough to say that we did not really mean what we said? Why does it burn my tongue to say that global warming is a fact whether you like it or not?" (Latour 2004: 224).
14. An example of an attempt to construct a realist account of quantum theory is David Bohm's hidden variable theory. Bohm argued that one key motivation for making his proposal was to show that such theories were in principle possible. He wrote "... it should be kept in mind that before this proposal was made there had existed the widespread impression that no conceptions of hidden variables at all, not even if they were abstract, and hypothetical, could possibly be consistent with the quantum theory". (Bohm 1980: 81). In his theory Bohm made a distinction between what he called implicate and explicate orders which he characterized as follows:

    > In the enfolded [or implicate] order, space and time are no longer the dominant factors determining the relationships of dependence or independence of different elements. Rather, an entirely different sort of basic connection of elements is possible, from which our ordinary notions of space and time, along with those of separately existent material particles, are abstracted as forms derived from the deeper order. These ordinary notions in fact appear in what is called the "explicate" or "unfolded" order, which is a special and distinguished form contained within the general totality of all the implicate orders. (Bohm 1980: xv)

He used the term 'unfoldment' to describe the process by which the implicate order becomes relevantly elevated (or "relevated") into the explicate order. He likens unfoldment to the way a television signal is decoded to produce a visible image on the screen. The implicate order in this analogy is the signal, screen, and television electronics, while the explicate order is the image produced. Another analogy deployed by him is the pattern produced by making small cuts in a folded paper that is then unfolded. The result is that a single original cut in the folded paper can give rise to widely correlated and separated elements of the patterns in the unfolded paper. In this case the explicate order is the unfolded pattern, and the cuts in the folded paper constitute the implicate order. These analogies, and the distinction between the implicate and explicate orders, allow him to deal with the problem of quantum entanglement where there are correlations between observables of entities separated by great distances in the explicate order.

15. Fuller notes that the Nobel Laureate, Steven Weinberg, did not hesitate to charge that Werner Heisenberg was "out of his depth," when he turned to pronouncing on the subjectivist epistemological implications of quantum mechanics, and when the physicist Alan Sokal dismissed the parallels noticed by Bohr between the complementarity principle and the metaphysical ideas of Daoism and Buddhism. Even Thomas Kuhn compiling oral histories of Bohr just before he died attributed his attempts to extend complementarity beyond physics to the mental infirmity of Bohr's advanced age (Fuller 2006: 118).
16. However we have seen that this was not the case with many of the leading founders of the quantum revolution. See Marin (2009) and Wilber (2001).
17. The relationship between science and religion has been the focus of an ongoing and extensive debate ever since the rise of modern science. There are some scholars such as Stephen Jay Gould who see science and religion as occupying separate "non-overlapping *magisteria*," but others including John Lennox, Thomas Berry, Brian Swimme, and Ken Wilber, see them as engaged in mutually fructifying dialogue. Perhaps the most systematic attempt to develop a taxonomy of these exchanges is Ian Barbour's categorization of these relations in terms of conflict, independence, dialogue, and integration. See Barbour (1997).
18. See Wigner (1967), pp. 153–184. This is part of a broader position associated with what has been labeled "quantum mysticism" by Marin (2009). Other writings of this genre include Zukav (1984), Talbot (2011), Wilber (2001), and Capra (1975). For a criticism of quantum mysticism, see Nanda (2003), Shermer (2005), and Stenger (1995).
19. This is also underscored by Abraham Pais in his biographical study *Niels Bohr's Times: In Physics, Philosophy, and Polity*:

> There have been those who attempted to portray him as a mystic. Bohr summarily dismissed such opinions. He believed he could express his views "without risking being misunderstood that it should be the purpose to introduce a mysticism which is alien to the spirit of natural science". (Pais 1991: 446)

Similarly Juan Miguel Marin emphasizes that Bohr maintained that it makes no difference whether the observer is a man, an animal, or a piece of apparatus. He argued that although Bohr saw quantum processes as evolving without conscious observers, he did not preclude the possibility that understanding consciousness may require extending quantum theory to include laws that went beyond physics (Marin 2009: 808–809).

Moreover, the significance Bohr attached to Eastern epistemological views also need not imply that he saw philosophy in general as important for his concerns. Pais makes this point by noting:

> [W]hen asked what kind of contributions he thought people like Spinoza, Hume, and Kant had made he replied evasively that that was difficult to answer … [although] Bohr did refer with great respect to Buddha and Lao Tse, however. (Pais 1991: 424)

20. The idea is also similar to the distinction made in medieval philosophy between natural religion and revealed religion. Generally in the West the notion of revealed religion as superior to the natural has been commonplace; in the East these valuations are reversed (with the *proviso* that what is natural includes mystical experience).
21. This is emphasized by Alvares (1992: 152–158).
22. Sal Restivo argues that these parallels are not merely contingently spurious but are socially constructed. He writes:

> Parallelist arguments rest on mystified versions of mysticism, and truncated models of physics (and science) as a social activity and process. This is related to the tendency in parallelism to draw together the mythical worlds of the mystic who experiences 'pure consciousness', and the scientist who experiences 'pure knowledge'. (Restivo 1982: 53)

23. Pioneers in popularizing this movement are Capra (1975), Zukav (1984), and Wilber (2001). There is even an interpretation of quantum properties as mind-created in Nobel physicist Eugene Wigner's view that consciousness collapses the wave function. See Wigner (1967: 153–184). Stenger (1995) has criticized the notion that human consciousness in the act of observation determines the outcome of what is observed—that mind creates reality. He argues that this has led to the misleading pantheistic mystical

conclusion that quantum physics supports the notion that in some ways our minds shape the universe as a whole.

24. Capra (1975). This study was anticipated by Siu (1964). For a critique of Capra and Zukav, and views similar to theirs, see Scerri (1989). Others have criticized Capra for building his case for physics–mysticism parallels on the outdated bootstrap model of strong-force interactions, which has now come to be replaced by the Standard Model. See Woit (2006: 141–145).

    However, Restivo argues that these parallels may simply be a matter of coincidence because science, whose knowledge is ever evolving, has currently come to bear spurious thematic parallels to some aspects of Eastern thought. He writes:

    > [T]he parallels between modern physics and Eastern mysticism may be spurious for reasons ranging from semantics to ideology. Evidence and logic do not support the idea that mystics have anticipated modern physicists. The fact that it is possible to identify parallels, analogies, and convergence between modern physics and mysticism may reflect a temporary condition in modern physical theory. Physics, and science in general, can be expected to develop and change in fundamental ways under the imperative of pressing inquiry ever forward. Mysticism, by contrast, appears to have arrived at certain 'ultimate' experiences and truths, at least within the boundaries of current human biological potentials, under a soteriological imperative that subordinates, ignores, and subverts active, open-ended inquiry. This does not mean that parallelism must be firmly and finally rejected as a manifestation of substantive linkages between physics and mysticism. (Restivo 1978: 167–168)

25. For a recent discussion of these issues linking science and Buddhist thought see Zajonc (2004) which records a dialogue between leading physicists, historians, and religious thinkers on current thought concerning linkages between quantum physics and Buddhist philosophy. See also Wallace (2003).
26. The debates initiated by Gross and Levitt continue to be relevant today in the field of science studies. See Dasgupta (2014) and Oreskes (2013). Surprisingly, Gross and Levitt completely ignore New Age thinkers. Perhaps they consider them to be beyond the pale of being even worthy of critique.
27. Gross and Levitt perceive these anti-science critics to be located in the humanities and social sciences:

    > Our subject is the peculiarly troubled relationship between the natural sciences and a large and influential segment of the American academic community which, for convenience but with great misgiv-

> ing, we call here "the academic left" … The category is comprised, in the main, of humanists and social scientists; rarely do the working natural scientists (who may nevertheless associate themselves with liberal or leftist ideas) show up within its ranks. (Gross and Levitt 1994: 3–4)

They then continue to divide the academic left into more specific categories inspired by postmodern, environmental, feminist, and multicultural perspectives:

> Postmodernism is grounded in the assumption that the ideological system sustaining the cultural and material practices of the Western European civilization is bankrupt and on the point of collapse. It claims that the intellectual schemata of the Enlightenment have been abraded by history to the point that nothing but a skeleton remains, held together by unreflective habit, incapable of accommodating the creative impulses of the future.
>
> The radical feminist view that science, like every other intellectual structure of modern society, is poisoned and corrupted by an ineradicable gender bias, is another vitally important element. An analogous accusation comes from multiculturalists, who view "Western" science as inherently inaccurate and incomplete by virtue of its failure to incorporate the full range of cultural perspectives. A certain strain of radical environmentalism condemns science as embodying the instrumentalism and alienation from direct experience of nature which are the twin sources of an eventual (or imminent) ecological doomsday.
>
> These ideas are the chief elements alloyed to form the academic left's challenge to conventional scientific thinking. It must be noted, however, that there is no canonical way of combining them … Rather it is as a congeries of different doctrines. With no well-defined center, each of which draws upon the notions we have cited in an idiosyncratic way, elaborating some of them with enthusiasm while leaving others in the background and rejecting still others completely. What enables them to coexist congenially, in spite of gross logical inconsistencies, is a shared sense of injury, resentment, and indignation against modern science. (Gross and Levitt 1994: 4–5)

Significantly, although Gross and Levitt have characterized the science wars mainly as a contest between sundry postmodern, environmental, feminist and multicultural thinkers from the humanities and social sciences arrayed against natural scientists, they also see these critics as inspired by the misplaced views of the complementarity interpretation of quantum theory endorsed by many pioneering scientists who created it.

28. The Copenhagen interpretation is the most widely accepted interpretation of quantum mechanics among physicists. This interpretation combines Bohr's complementarity principle with Heisenberg's uncertainty relations. Perhaps second to it in popularity is the many-worlds interpretation, although there are many other alternative interpretations, albeit winning acceptance only among pockets of followers. These include quantum logic, de Broglie-Bohm theory, von Neumann interpretation, many-minds interpretation, objective collapse theories, and even Max Born's ensemble interpretation.
29. It is noteworthy that even Einstein, who was sympathetic to Bohm's search for a realistic alternative to the Copenhagen interpretation, nevertheless repudiated his proposed interpretation. In a letter to Max Born he writes:

    > Have you noticed that Bohm believes (as de Broglie did, by the way, 25 years ago) that he is able to interpret the quantum theory in deterministic terms? That way seems too cheap to me. (Letter of 12 May 1952 from Einstein to Max Born, in *The Born–Einstein Letters*, Macmillan, 1971, p. 192)

30. These papers triggered the so-called Sokal affair or Sokal hoax which led to questions about the scholarly standards of journals in the social science and humanities concerned with science studies. It also promoted concerns that postmodern philosophy had depleted traditional standards of rigor and integrity in such disciplines. It came to an end only after a similar scandal developed in the natural sciences following the Bogdanoff affair—sometimes referred to as the reverse-Sokal controversy—which in turn made questionable standards of peer-review in some leading international journals of physics as well.
31. This is evident when Sokal writes that the complementarity principle is by no means universally accepted by physicists, that even many who do accept it do so because of the immense prestige of Bohr rather than its helpfulness in better understanding quantum theory (Sokal 2008: 14), and that physicists in general have accepted the Copenhagen interpretation of Bohr and Heisenberg "in a rather dogmatic way" (ibid. p. 201). However, though he no longer blames Bohr for the postmodern exaggerations of the Copenhagen interpretation of quantum theory he does note that "postmodernist musings on quantum mechanics ... [exhibit] a fondness for the most subjectivist writings of Heisenberg and Bohr, interpreted in a radical way that goes far beyond their own views (which are in turn vigorously disputed by many physicists and philosophers of science)" (ibid. p. 12).
32. It is also important to note that the traditions of science seen as parts of a postmodern science—ecology, chaos theory, and quantum physics—are

proposing objective construction of phenomena as shaped by high sensitivity to environmental contexts, but not embracing relativist social or cultural constructions. This distinction between the postmodern social constructivism and postmodern sciences has been noted by Brown:

> There is an interesting distinction to be made between postmodern accounts of science (which are invariably anti-objectivist) and postmodern science. With surprising frequency postmoderns cheerfully embrace a number of particular sciences. Quantum mechanics (or at least aspects of it) and chaos are favorites. It is easy to see why. Heisenberg's uncertainty principle (according to a common interpretation) suggests that we humans are not merely observing the world in a passive way, but are somehow actively involved in making it what it is. Chaos theory makes the world out to be a wildly unpredictable and uncontrollable place. These feed into popular postmodern themes about the subjectivity, contingency, instability and complexity of the world. (Brown 2001: 94)

Brown's analysis explains why there are affinities between postmodern thought and the postmodern sciences but it is also important to note the differences. Postmoderns emphasize the role of the social or linguistic context as the paramount factor that shapes what we observe. But quantum properties are not an outcome of making an observation in a social or linguistic context but a physical environmental context. Similarly chaos theory is the result of sensitivity of phenomena to environmental contexts in which they evolve and has nothing to do with the postmodern notion that sociocultural contexts shape knowledge.

33. John Baez (2010) was one of the earliest to draw a comparison between the Sokal and Bogdanoff affairs. It is noteworthy that the main papers by Grichka and Igor Bogdanoff are closely connected with the applications of quantum theory. They include the following:

> Grichka Bogdanoff and Igor Bogdanoff, "Topological Field Theory of the Initial Singularity of Space-Time," *Classical and Quantum Gravity* (2001) **18**: 4341–4372; Grichka Bogdanoff and Igor Bogdanoff, "Space-Time Metric and the KMS Condition at the Planck Scale," *Annals of Physics*, (2002) **296:** 90–97; Grichka Bogdanoff and Igor Bogdanoff, "KMS Space-Time at the Planck scale," *Nuovo Cimento*, (2002) **117B**: 417–424; Igor Bogdanoff, "Topological Origin of Inertia," *Czechoslovak Journal of Physics*, (2001) **51:** 1153–1236; and Igor Bogdanoff, "The KMS State of Space-Time at the Planck Scale," *Chinese Journal of Physics*, (2002) **40:** 149–158.

34. See Plotnitsky 1994, pp. 249–260 for a more detailed discussion.
35. We also have a few studies that give more extended attention to Bohr's epistemological views beyond physics. Particularly synoptic is the section "Complementarism" in Abraham Pais' (1991: 438–447) biographical study of Bohr. Other works that look at Bohr's attempts to extend complementarity into the other sciences include Bohr (1999), Folse (1985), and Faye and Folse (1994).
36. Quoted in Pais (1991), p. 439.
37. Wertheimer (1924), p. 10. He writes:

    > It is our task to inquire whether a logic is possible which is *not* piecemeal. Indeed the same question arises in mathematics also. Is it *necessary* that all mathematics be established upon a piecewise basis? What sort of mathematical system would it be in which this were *not* the case? There have been attempts to answer the latter question but almost always they have fallen back in the end upon the old procedures. This fate has overtaken many, for the result of training in piecewise thinking is extraordinarily tenacious. It is not enough and certainly does not constitute a solution of the principal problem if one shows that the axioms of mathematics are both piecemeal and at the same time evince something of the opposite character. The problem has been scientifically grasped only when an attack specifically designed to yield positive results has been launched. Just how this attack is to be made seems to many mathematicians a colossal problem, but perhaps the quantum theory will force the mathematicians to attack it.

    The recent discovery of fractal geometry perhaps points in the direction of the mathematics Wertheimer envisaged. It allows complex images and biological structures to result from the application of very simple mathematical transformations.
38. For a more recent study of these gestalt and quantum parallels see Chiara et al. (2006). They argue that quantum computation suggests a new holistic semantics that has characteristic features associated with perceptual gestalts, since quantum information in qubits has a global structure that determines the meanings of parts in a manner analogous to how gestalt percepts shape their parts.
39. Quoted in Pais (1991), p. 443.
40. The distinction between the genotype and the phenotype was first proposed by Wilhelm Johannsen in 1911 to clarify the difference between what an organism inherits and what its heredity produces. A similar distinction was proposed by August Weismann when he distinguished the germ plasm (heredity) and somatic cells (the body).

Contemporary biologists take the genotype of an organism to be carried by the instructions in its genetic code. However organisms with the same genotype may not appear or behave in the same way because these are influenced both by environmental and developmental factors. For the same reason organisms that look alike may have developed from different genotypes.

41. Although originally formulated by the chemist James Lovelock, the Gaia hypothesis later came to be co-developed by the microbiologist, Lynn Margulis. In the beginning it was repudiated by the scientific community because its assumption that the physical conditions for supporting life on earth were created by life itself seemed excessively teleological. However, it now has wider acceptance, especially in disciplines such as systems ecology, Earth system science, biogeochemistry, and geophysiology. See Lovelock (1982) and Margulis and Hinkle (1988).
42. See Dickinson (1987).
43. Henry Folse argues that Bohr saw the complementarity viewpoint as capable of resolving controversies surrounding teleological explanations in biology. He maintains that Bohr's position can be connected to three stages in these debates: First the debates between the mechanical and vitalist positions that confronted Bohr and led him to reject ontologically teleological explanations; second the positivist position that the controversy was a pseudo-problem that Bohr also rejected; and finally the current "autonomist" versus "provincialist" debate. Folse argues that Bohr's viewpoint can contribute to the latter because it presents an interactionist ontology that defends the irreducibility of functionally teleological explanations. See Folse (1990).

    The biophysicist Max Delbruck, who received the Nobel Prize for his discovery that bacteria developed resistance to viruses as a result of advantageous genetic mutations, was originally inspired by an attempt to extend Bohr's complementarity viewpoint into the biological sciences. See Roll-Hansen (2000). A more recent attempt to extend such connections can be found in Theise and Kafatos (2013).
44. The *Taijitu* is now the symbol for the Chinese religion of Daoism, but it is also often used by non-Daoists to represent the notion of the harmonious co-existence of opposites. Surprisingly patterns similar to it can be found in Roman, as well as Celtic and Etruscan, iconography. Indeed the *Notitia Dignitatum*, a fifth-century document of the Roman government, contains some of the earliest known representations of what has since come to be known as the yin and yang symbols. These Roman patterns antedate by almost seven hundred years the earliest Daoist versions. See Monastra (2000).

45. For the importance of the clockwork metaphor in modern science see Dijksterhuis (1969). Although Bacon, Descartes, and Hooke espoused a mechanical conception, closely linked to the idea that the universe ran like a clock, this metaphor came to be consolidated only after the Newtonian synthesis.
46. This has been emphasized by Dijksterhuis who writes:

    > As a matter of fact, he (Descartes) states explicitly that between natural bodies and artifacts produced by skilful artisans he recognizes no other difference than one of size: that which takes place invisibly in the former, in the latter happens on so big a scale that we can observe it. For the rest there is not a single difference between a running clockwork and a growing tree. That is also why those who are versed in the construction of automata are best fitted to guess the true process of natural phenomena, the mechanisms hidden in them. (Dijksterhuis 1969: 415)

47. In addition, we will find that acknowledging the distinctive structure of grown properties, and seeing complementarity epistemology as a way of coming to terms with them, also has additional advantages. It provides a basis to articulate many environmental, multicultural, and ecofeminist critiques of mechanical science, and their embrace of the context sensitivity of quantum science, without the expedient of invoking postmodern constructivism or New Age idealism. Indeed we will find that complementarity offers an alternative framework for such critiques that makes possible an objective science that is neither postmodern relativist nor New Age idealist. Ontologically this has parallels to the ethics debates addressed by Sterba (2001).

Bibliography

Asimov, Isaac. 1979. Do Scientists Believe in God? *Gallery* 7(7): 51–52 and 104–106.

Allinson, Robert E. 1998. Complementarity as a Model for East-West Integrative Philosophy. *Journal of Chinese Philosophy* 25: 505–517.

Alvares, Claude. 1992. *Science, Development and Violence: The Revolt Against Modernity*. Delhi, New York: Oxford University Press.

Baez, John. 2010. *The Bogdanoff Affair* (22 Oct 2010). http://math.ucr.edu/home/baez/bogdanoff/. Accessed 1 Sept 2015.

Barad, Karen. 2007. *Meeting the Universe Halfway: Quantum Physics and the Entanglement of Matter and Meaning*. Durham: Duke University Press.

Barbour, Ian. 1997. *Religion and Science: Historical and Contemporary Issues*. San Francisco: Harper One.

Beckwith, Christopher I. 2015. *Greek Buddha: Pyrrho's Encounter with Early Buddhism in Central Asia*. Oxford: Princeton University Press.

Beller, Mara. 1998. The Sokal Hoax: At Whom are We Laughing? *Physics Today* 51: 29–34.

———. 1999. *Quantum Dialogue: The Making of a Revolution*. Chicago, IL: University of Chicago Press.

Birkhoff, Garrett, and John Von Neumann. 1936. The Logic of Quantum Mechanics. *Annals of Mathematics* 37(4): 823–843.

Bohm, David. 1980. *Wholeness and the Implicate Order*. London: Routledge.

Bohr, Niels. 1928. The Quantum Postulate and the Recent Development of Atomic Theory, *Nature* 121(Suppl.): 580–590. Republished in *Niels Bohr Collected Works* (1985) Vol. 6 *Foundations of Quantum Physics I (1926–1932)* pp. 148–158.

———. 1949. Discussion with Einstein on Epistemological Problems in Atomic Physics. In Schilpp (1949) pp. 201–241.

———. 1950. On the Notions of Causality and Complementarity. *Science, New Series* 111(2873): 51–54.

———. 1958. *Atomic Physics and Human Knowledge*. New York: John Wiley & Sons.

———. 1999. *Collected Works Vol. 10: Complementarity Beyond Physics (1928–1962)*, ed. David Favrholdt. Elsevier.

Brown, James Robert. 2001. *Who Rules in Science: An Opinionated Guide to the Wars*. Cambridge, MA: Harvard University Press.

Bruton, Eric. 1989. *The History of Clocks and Watches*. London: McDonald & Co.

Capra, Fritjof. 1975. *The Tao of Physics: An Exploration of the Parallels Between Modern Physics and Easter Mysticism*. Berkeley, CA: Shambala.

Chiara, Maria Luisa Dalla, Roberto Giuntini, and Giuliano Toraldo di Francia. 2006. Holistic Quantum Computational Semantics and Gestalt-Thinking. In *AIP Conference Proceedings*, vol. 844, 86–100.

Dasgupta, Subrata. 2014. Science Studies Sans Science: Two Cautionary Postcolonial Tales. *Social Scientist* 42(5/6): 43–61.

Descartes, René. 1982. *Principles of Philosophy*: Translated, with Explanatory Notes by V.R. Miller and R.P. Miller. Dordrecht: Kluwer Academic Publishers.

Dickinson, Robert E. (ed.) (1987) *The Geophysiology of Amazonia: Vegetation and Climate Interactions*. New York: Wiley-Interscience.

Dijksterhuis, E.J. 1969. *The Mechanization of the World Picture*. London: Oxford University Press.

Dummett, M. 1976. Is logic empirical? In *Contemporary British Philosophy, 4th series*, ed. H. D. Lewis, 45–68 London: Allen and Unwin. Reprinted in M. Dummett, *Truth and other Enigmas* (London: Duckworth,1978), pp. 269–289.

Dumoulin, Heinrich. 1988, 1990. *Zen Buddhism: A History*, vol. 2. Trans. James W. Heisig and Paul Knitter. New York: Macmillan.

Faye, Jan, and Henry J. Folse, eds. 1994. *Niels Bohr and Contemporary Philosophy*. Boston: Kluwer Academic Publishers.

Folse, Henry J. 1985. *The Philosophy of Niels Bohr: The Framework of Complementarity*. Amsterdam: Elsevier Science Publishers.

Folse, Henry J 1990. Complementarity and the Description of Nature in Biological Science. *Biology and Philosophy* 5(2): 211–224.

Fuller, Steve. 2006. *The Philosophy of Science and Technology Studies*. New York: Routledge.

Gilbert, Scott. 2010. *Developmental Biology*. Sunderland, MA: Sinauer Associates Inc.

Goonatilake, Susantha. 1982. *Crippled Minds: An Exploration into Colonial Culture*. New Delhi: Vikas Publishing House.

Gross, Paul R., and Norman Levitt. 1994. *Higher Superstition: The Academic Left and Its Quarrels with Science*. Baltimore, MD: The Johns Hopkins University Press.

Hacking, Ian. 1999. *The Social Construction of What?* Cambridge, MA: Harvard University Press.

Heisenberg, Werner. 1963. *Physics and Philosophy*. London: Allen & Unwin.

Hooker, Clifford. 1972. The Nature of Quantum Mechanical Reality: Einstein Versus Bohr. In *Paradigms and Paradoxes: The Philosophical Challenge of the Quantum Domain*, ed. Robert G. Colodny, 67–302. Pittsburgh, PA: University of Pittsburgh.

Jayatilleke, K.N. 1967. The Logic of Four Alternatives. *Philosophy East and West* 17: 1–4. Hawaii: University of Hawaii Press.

Katsumori, Makoto. 2011. *Niels Bohr's Complementarity: Its Structure, History, and Intersections with Hermeneutics and Deconstruction*. Berlin: Springer.

Köhler, Wolfgang. 1959. Gestalt Psychology Today. *American Psychologist* 14: 727–734. Republished in *Classics in the History of Psychology*. See http://psych-classics.yorku.ca/Kohler/today.htm. Accessed 1 Sept 2015.

Kuhn, Thomas. 1970. *The Structure of Scientific Revolutions*. Chicago: University of Chicago Press.

Kuzminski, Adrian. 2008. *Pyrrhonism: How the Ancient Greeks Reinvented Buddhism*. New York: Lexington Books.

Latour, Bruno. 2004. Why Has Critique Run Out of Steam? From Matters of Fact to Matters of Concern. *Critical Inquiry* 30(2): 225–248.

Lovelock, James. 1982. *Gaia: A New Look at Life on Earth*. Oxford: Oxford University Press.

Loy, David. 1988. *Nonduality: A Study in Comparative Philosophy*. New Haven & London: Yale University Press.

Margulis, Lynn and Gregory Hinkle. 1988. Biota and Gaia. *Abstracts of First Chapman Conference on Gaia Hypothesis.*

Marin, Juan Miguel. 2009. 'Mysticism' in Quantum Mechanics: The Forgotten Controversy. *European Journal of Physics* 30: 807–822.

Monastra, Giovanni. 2000. The "Yin-Yang" Among the Insignia of the Roman Empire? *Sophia* 6(2). See http://www.estovest.net/tradizione/yinyang_en.html. Accessed 1 Sept 2015.

Nakamura, Hajime. 1975. *Parallel Developments: A Comparative History of Ideas.* Tokyo and New York: Kodansha.

Nanda, Meera. 2003. *Prophets Facing Backward: Postmodern Critiques of Science and Hindu Nationalism in India.* New Jersey: Rutgers University Press.

Oppenheimer, J. Robert. 1954. *Science and the Common Understanding.* Oxford: Oxford University Press.

Oreskes, Naomi. 2013. On the "Reality" and Reality of Anthropogenic Climate Change. *Climatic Change* 119(3–4): 559–560.

Pais, Abraham. 1991. *Niel's Bohr's Times: In Physics, Philosophy, and Polity.* Oxford: Clarendon Press.

Plotnitsky, Arkady. 1994. *Complementarity: Anti-Epistemology After Bohr and Derrida.* Durham: Duke University Press.

Putnam, Hilary. 1968. Is Logic Empirical? In *Boston Studies in the Philosophy of Science*, eds. Dordrecht: D. Reidel, vol. 5, 216–241. Reprinted as "The Logic of Quantum Mechanics" in Putnam's *Mathematics, Matter and Method* (1979) Cambridge University Press. pp. 174–197.

Reichenbach, Hans. 1944. *The Philosophical Foundations of Quantum Mechanics.* University of California Press. Reprinted by Dover (1998).

Restivo, Sal P. 1978. Parallels and Paradoxes in Modern Physics and Eastern Mysticism: I—A Critical Reconnaisance. *Social Studies of Science* 8: 143–181.

Restivo, Sal P 1982. Parallels and Paradoxes in Modern Physics and Eastern Mysticism: II—A Sociological Perspective on Parallelism. *Social Studies of Science* 12: 37–71.

Roll-Hansen, Nils. 2000. The Application of Complementarity to Biology: From Niels Bohr to Max Delbrück. *Historical Studies in the Physical and Biological Sciences* 30(2): 417–442.

Rosenfeld, L. 1963. Neils Bohr's Contribution to Epistemology. *Physics Today* 16(10): 47.

Rosenfeld, Leon. 1996. Complementarity: Bedrock of the Quantal Description. In *Foundations of Quantum Physics II (1933–1958) Niels Bohr Collected Works*, vol 7, ed. Kalckar J. Joegen. Amsterdam: Elsevier.

Ross, W.D. 2006. *Aristotle's Metaphysics: A Revised Text.* Translated with Introduction and Commentary, vol. 2.

Sarukkai, Sundar. 2005. *Indian Philosophy and Philosophy of Science*. New Delhi: Centre for Studies in Civilizations.

Scerri, E.R. 1989. Eastern Mysticism and the Alleged Parallels with Physics. *American Journal of Physics* 57(8): 687–692.

Shermer, Michael. 2005. Quantum Quackery. *Scientific American* 292(1): 34.

Siu, R.G.H. 1964. *The Tao of Science: An Essay on Western Knowledge and Eastern Wisdom*. Cambridge, MA: The MIT Press.

Smart, Ninian. 2000. *World Philosophies*. London & New York: Routledge.

Snow, C.P. 1960. *The Two Cultures and the Scientific Revolution*. Cambridge, MA: Cambridge University Press.

Sokal, Alan. 1996a. Transgressing the Boundaries: Toward a Transformative Hermeneutics of Quantum Gravity. *Social Text* 46/47: 217–252.

———. 1996b. A Physicist Experiments With Cultural Studies. *Lingua Franca* 6(4): 62–64.

———. 2008. *Beyond the Hoax: Science, Philosophy, and Culture*. Oxford: Oxford University Press.

Sokal, Alan, and Jean Bricmont. 1998. *Fashionable Nonsense: Postmodern Intellectuals' Abuse of Science*. New York: Picador.

———. 1999. Letters: Sokalratic Debate Continues, Fueled by Latour and Copenhagen Interpretations. *Physics Today* 52(8): 15 & 82.

Stenger, Victor J. 1995. *The Unconscious Quantum: Metaphysics in Modern Physics and Cosmology*. Amherst, NY: Prometheus Books.

Sterba, James P. 2001. *Three Challenges to Ethics: Environmentalism, Feminism, and Multiculturalism*. New York: Oxford University Press.

Talbot, Michael. 2011. *The Holographic Universe: The Revolutionary Theory of Reality*. New York: Harper Perennial.

Theise, Neil, and Menas Kafatos. 2013. Complementarity in Biological Systems: A Complexity View. *Complexity* 18(6): 11. doi:10.1002/cplx.21453.

Wallace, B. Alan, ed. 2003. *Buddhism and Science: Breaking New Ground*. New York: Columbia University Press.

Weber, Renee. 1986. *Dialogues with Scientists and Sages: The Search for Unity*. London: Routledge and Kegan Paul.

Weinberg, Steven. 2001. *Facing Up: Science and Its Cultural Adversaries*. Cambridge, MA: Harvard University Press.

Wertheimer, Max. 1924. Gestalt Theory. In *A Source Book of Gestalt Psychology Prepared (1974)*, ed. Willis D. Ellis. London: Routledge & K. Paul.

Wheeler, John Archibald. 1963. No Fugitive and Cloistered Virtue—A tribute to Niels Bohr. *Physics Today* 16(1): 30–32.

Wigner, E.P. 1967. *Symmetries and Reflections*. Bloomington, IN: Indiana University Press.

Wilber, Ken, ed. 2001. *Quantum Questions: Mystical Writings of the World's Great Physicists.* Originally published 1984. Boulder, CO: Shambhala.
Woit, Peter. 2006. *Not Even Wrong: The Failure of String Theory and the Search for Unity in Physical Law.* New York: Basic Books.
Zajonc, Arthur, ed. 2004. *The New Physics and Cosmology: Dialogues with the Dalai Lama.* Oxford: Oxford University Press.
Zukav, Gary. 1984. *The Dancing Wu Li Masters: An Overview of the New Physics.* New York: Bantam.

CHAPTER 2

Biological Complementarity of the Molecular and Functional

2.1 Quantum Complementarity and Grown Properties

This chapter attempts to develop and defend Bohr's extension of complementarity to the biological arena by combining the molecular and functional, or genetic and environmental, explanations as necessary to achieve a more comprehensive understanding of phenomena. However, before we attempt to understand biological complementarity let us examine why complementarity has come to play such an important role in atomic physics—a domain far removed from the biological world. Surprisingly, complementarity in atomic physics reflects the structural similarities of quantum properties to grown properties.

The hypothesis that quantum properties are grown explains why quantum theory has generated a great deal of philosophical interest and controversy ever since its formulation in 1926. Indeed, soon after the mathematical formalism of the theory had been consolidated there developed a long and continuing debate between Bohr and many leading physicists, who found the theory unpalatable.[1] Among those who rejected quantum theory were those who had made seminal contributions to the birth of the theory, such as Max Planck, Albert Einstein, Erwin Schrodinger, and Louis de Broglie. These leading scientists questioned the theory's adequacy to represent all aspects of atomic phenomena.

The most articulate dissenting voice was Einstein's. His vehement objections to Bohr's complementarity perspective led to a highly publicized

Arun Bala, *Complementarity Beyond Physics*,
DOI 10.1007/978-3-319-39784-9_2

epistemological debate concerning the nature and objectives of scientific investigation. Einstein maintained that the epistemology of complementarity masked the limitations of quantum theory to fully explain atomic events. In particular, Einstein tried to demonstrate that quantum theory was incomplete because atomic objects simultaneously possess those properties, such as a precise position and momentum, for example, which Bohr treats as mutually exclusive and complementary.

However, Bohr contended that quantum theory precluded the simultaneous existence of a precise position and a precise momentum for an atomic particle because such properties cannot simultaneously exist in nature. Hence, Bohr countered that quantum theory provided the best account of atomic phenomena *possible*, and that Einstein was laying down preconditions for what nature should be like over and above what was warranted by experiment, observation, and quantum theory. Bohr saw what Einstein perceived as limitations of quantum theory—namely, its inability to predict simultaneously properties represented by non-commutating operators, such as position and momentum, as actually providing a better representation of nature because these properties cannot have precise values at the same time. Hence, he charged Einstein with demanding more from quantum theory than nature rendered possible.[2]

Einstein's most widely discussed paper against quantum theory was written in 1935 in collaboration with Boris Podolsky and Nathan Rosen.[3] In it, the writers proposed an experiment that brought to the forefront the problem of the epistemic status of quantum properties, such as position and momentum, attributed to atomic systems. In their paper Einstein, Podolsky, and Rosen (henceforth referred to as EPR) consider the case of a single particle that decomposes into two equal particles A and B traveling in opposite directions. They argue that, given the law of conservation of momentum, each particle would have a momentum equal and opposite to the other. Hence, by deploying instruments to measure the momentum of A, we can predict with certainty the momentum of B prior to its measurement. It would be equal in magnitude, but opposite in direction, to the momentum of A. Similarly, if we were to deploy an instrument to measure the position of A, we can predict the position of B precisely. The position of B would be in the opposite direction to A, and at an equal distance from the position where the particles separated. Thus, by measuring either the position or momentum of A, we can predict the position or momentum that would be measured for B.

EPR then continue their argument as follows. They begin by assuming that it is reasonable to suppose that once the particles have separated any measurement on A has no influence on B. Moreover, since we are free to choose whether we want to measure the momentum or position of A, *and* whatever property we choose allows us to predict the corresponding property for B, we must conclude that B possesses this property before the measurement is made. This proves, according to EPR, that B has both a well-defined position and a well-defined momentum even prior to measurement. But quantum theory is unable to specify with certainty both the position and momentum of the particle B at the same time. Moreover, the principle of uncertainty, which is a logical consequence of quantum theory, also precludes any simultaneous assignation of a precise momentum and a precise position to B. Since their argument shows, so say EPR, that B actually has both a simultaneously well-defined momentum and position, we have to conclude that quantum theory is incomplete because it cannot specify completely properties that actually co-exist in nature.

EPR recognize that their conclusion follows only if we make two assumptions—assumptions they consider so unproblematic as to be acceptable to most physicists. They formulate these assumptions in the form of two principles—the principle of reality and the principle of locality. The reality principle is expressed by them as follows:

> If, without in any way disturbing a system, we can predict with certainty (i.e., with probability equal to unity) the value of a physical quantity, then there exists an element of physical reality corresponding to this physical quantity. (Einstein et al. 1935: 777)

In the example of the particle B considered above, its position and momentum can be predicted with certainty without disturbing it—we can predict those values by doing our measurements on A. Hence, according to EPR, both the position and momentum of B have an element of physical reality.

This shows, according to EPR, that quantum theory is incomplete. They reason as follows. For a physical theory to be complete it must be such that "every element of the physical reality must have a counterpart in the theory." (Einstein et al. 1935: 777) Since quantum theory cannot attribute both a well-defined position and a well-defined momentum to a physical system that has both these properties at the same time (as appears to be the case with the particle B), it is an incomplete theory.

Of course, such a conclusion would fail to follow if a measurement made on A is considered to disturb B even when A is widely separated from B. However, EPR consider this unlikely. To make this assumption explicit they propose a second principle, the principle of locality, which they think most physicists would be prepared to accept. The principle essentially assumes that there is no interaction or influence between two systems, such as A and B, which are separated sufficiently far from each other even if they have interacted in the past. This principle of locality is expressed by EPR as follows:

> [If] at the time of measurement the two systems no longer interact, no real change can take place in the second system in consequence of anything that may be done to the first system. (Einstein et al. 1935: 779)

They argue that the particles A and B can be reasonably expected to satisfy this condition. Hence, EPR conclude, if we accept the reality and locality principles, that they deem we cannot reasonably reject, then quantum theory is incomplete—its formalism fails to represent all elements of physical reality, such as the position and momentum of B at the same time.

The EPR argument assumed an added significance when the physicist John Bell, in 1964, nearly 30 years after it was first formulated, demonstrated mathematically that the EPR "local realistic" constraints on a physical theory would imply testable constraints on the empirical predictions of the theory. This transformed the debate between Einstein and Bohr because it now became possible to experimentally test whether actual observational data conform to the assumptions invoked by EPR. In fact Bell developed a set of so-called "Bell Inequalities" in the form of precise mathematical formulae that should be satisfied by any system that realized the constraints of local realism proposed by EPR. Moreover, he demonstrated that quantum theory was incompatible with local realism because the empirical predictions of the theory would violate these inequalities.[4]

Indeed the numerous experiments that have been performed since Bell proposed his test confirm the predictions of quantum theory but violate the Bell inequalities.[5] It seems to follow that quantum theory requires us to relinquish local realism.

The experiments themselves are so intriguing and their implications so central to the interpretation we wish to propose—that is, quantum properties are grown properties—that we shall examine them in some detail. However, these experiments are not designed to measure the position or momentum of a microsystem, because it has not been possible to develop experiments that use these particular parameters to test the assumptions in local realism. Instead the early experiments were designed to measure photon polarization—another quantum property that also exhibits correlations of the sort found between position and momentum in the particles A and B considered by EPR. Since then many different experiments have been conducted involving different techniques to measure quantum correlations to test the Bell inequalities.

Let us consider one of the earliest experiments to test the Bell inequalities conducted by Alain Aspect, Philippe Grangier, Gerard Roger and Jean Dalibard.[6] Using photon polarization correlations, it brings out all the essential aspects of such Bell tests of quantum theory. It also serves to illustrate the utility of the hypothesis that quantum properties are grown. Even though these experiments were conducted more than 30 years ago, they continue to reflect the basic structure of all Bell test experiments to determine whether the real world conforms to local realism. Their use of photon polarization properties also came to be adopted in many later experiments. This contrasts with the original Bell paper where the envisaged experiment was designed to measure the position or momentum of atoms. Indeed many of the experiments that followed were largely designed to refine these early experiments to avoid what has been characterized as the locality and detection loopholes. Eliminating the locality loophole in an experiment involved ensuring that each separate measurement in each wing of the experiment allows a new setting to be chosen and the measurement to be completed before signals could communicate information about the settings from one wing of the experiment to the other. Freeing the experiment from the detection loophole involves ensuring that nearly 100% of measurements in one wing are co-related with a successful measurement in the other wing. Combining high efficiency with rapid implementation of measurement settings has been the major target of Bell test experiments over the last 30 years.[7]

The following diagram (Fig 2.1) is a schematic representation of the Aspect, Grangier, Roger and Dalibard experiment to test Bell inequalities.[8] The experiment uses an excited gas that emits photons in pairs such that each member of a pair flies off in opposite directions. The polarization of any photon can be experimentally measured by means of polarization detectors placed on opposite sides of the source as shown in the diagram. These measurements yield results that will be either positive (+) or negative (−). The design of the apparatus allows the polarizers to rotate so that

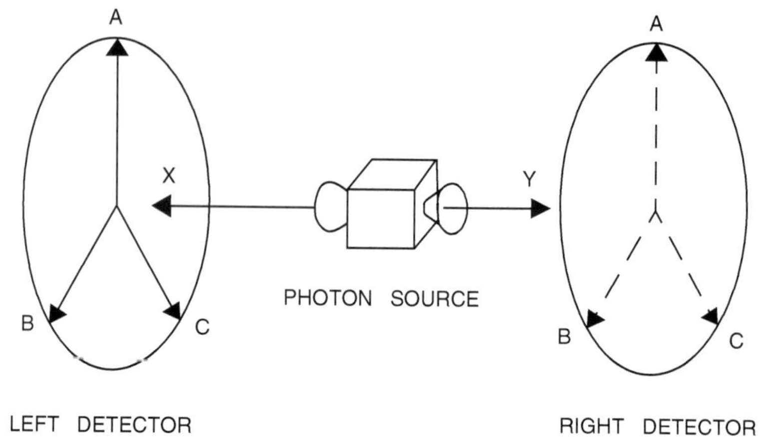

Fig. 2.1 Schematic Diagram of Photon Polarization Experiment

they can be inclined in three different directions A, B, or C that are 120 degrees apart. Thus when they are both at A (i.e. the A-A position) they are inclined parallel to each other. This would also be the case in the B-B or C-C positions.

The experiment shows that whenever the detectors are parallel in the A-A direction, the photon X that enters the left hand detector registers the same polarization as the photon Y—its corresponding member in the pair—that enters the right hand detector. Thus X and Y of any emitted pair register either both plus (+ +) or both minus (– –), but never (+–) or (– +).

We can therefore assume that the polarizations of the photons of any pair are correlated in the A direction. If the polarization of the photon X is found to be positive, we can predict with certainty that a measurement of its associated photon Y would also yield a polarization value that is positive. Hence, if we assume the EPR reality principle, we appear to be forced to conclude that the polarization of the photon Y has an element of physical reality even prior to its measurement in the A direction.

We can repeat the same experiment and the same argument for the B-B and C-C settings. Hence, accepting the EPR reality principle, we seem to be compelled to conclude that the photon Y has a well-defined polarization in all three directions even before a measurement takes place. Moreover, since we can switch the positions of X and Y, by measuring the polarization of Y in order to predict the polarization that would be measured on X, we have to conclude, so the EPR argument suggests, that X also has well-defined polarization values in all three directions.

It is now possible to use Bell's theorem to set a mathematical limit on the degree of polarization correlations between X and Y when the polarizers are inclined 120 degrees to each other, that is, in the A-B, A-C, B-A, B-C, C-A or C-B directions. Then the Bell theorem predicts that at least one third of the photon pairs should register the same sign (i.e. + or –) when the detectors are set in different directions.[9] However, actual experimental data reveal that only one quarter of the photon pairs register the same sign when the switches are in different settings. This violates the Bell inequalities but conforms completely to the predictions of quantum theory.

Of course, it is possible to explain the failure of the correlations to observe the Bell inequalities by assuming that some sort of interaction or disturbance travels from the first photon, say X, when its polarization is measured, to the second photon Y before its polarization is measured, so as to disturb the polarization of Y. After all, the Bell inequalities were computed by assuming the locality principle of EPR.

However, one can avoid the propagation of any such influence from one photon to the other by ensuring that no disturbance can reach Y from X in the interval between measuring the polarization of X and measuring that of Y, even if such a disturbance were to propagate at the velocity of light—the fastest speed at which material influences can travel according to the theory of relativity. We can do this by making the distance between the polarizers sufficiently large and determining the actual settings of the polarizers at the very last moment. Then the measurement interval can be made sufficiently small so that the observed correlations cannot be attributed to any mechanical disturbance emanating from X and reaching Y. Such a disturbance would violate the relativistic principle that no energetic transmission can occur at velocities exceeding that of light. Even after taking such precautions the experimental results contravened the Bell inequalities and confirmed the predictions of quantum theory.

The significance of these results actually extends far beyond vindicating quantum theory—it appears to refute all local realistic theories independent of the acceptability of quantum theory. James Cushing has appropriately described the Bell theorem as a "no-go" theorem which, in conjunction with the polarization experimental results, "refutes a whole category of (essentially) classical theories without ever mentioning quantum mechanics."[10] Hence, we seem to be forced to conclude that either the locality principle or the reality principle, or both, are violated in nature.

Surprisingly Bohr refuses to draw this conclusion. According to him the complementarity interpretation requires us to neither relinquish the reality principle nor the locality principle. First, he accepts the locality principle by stating quite categorically that there can be no question of any mechanical disturbance traveling from one system to the other. (Bohr 1935: 700) Thus he rules out any supraliminal connection between the two systems whose correlated properties are measured.

Second, Bohr does not reject Einstein's formulation of the reality principle. However, he argues that EPR wrongly abstract the property measured from the context of measurement by treating it as preexisting this context. Bohr argues that this is untenable, and that the measured property can only be defined relative to the particular experimental situation designed to measure it. In his reply to the EPR paper of 1935 he writes:

> My main purpose … is to emphasize that in the phenomena concerned we are not dealing with an incomplete description characterized by the arbitrary picking out of different elements of physical reality at the cost of sacrificing other such elements, but with a rational discrimination between essentially different experimental arrangements and procedures which are suited either for an unambiguous use of the idea of space location, or for a legitimate application of the conservation theorem of momentum. … Indeed we have in each experimental arrangement suited for the study of proper quantum phenomena not merely to do with an ignorance of the value of certain physical quantities, but with the impossibility of defining these quantities in an unambiguous way. (Bohr 1935: 699)

The point Bohr stresses is that quantum properties can only be defined within the context of well-defined experimental arrangements. Only then can we, as the above passage implies, make "unambiguous use" of classical concepts. Properties such as position and momentum, and we might add those such as polarization of a photon in different directions, cannot be simultaneously measured since the experimental arrangements that would enable us to do this cannot be simultaneously deployed. These experimental arrangements are mutually exclusive. To measure the position, or polarization in a particular direction, of an atomic object we need to set up an experimental situation that prevents us from also setting up the situation to measure momentum or polarization in a different direction. Hence position and momentum, and the polarizations in different directions, cannot be considered as simultaneously definable with certainty for a particle since they cannot be simultaneously measured.

Perhaps the most explicit formulation by Bohr of the complementarity view is the following[11]:

> [Quantum mechanics reveals that] it is most decisive to recognize that however far the phenomena transcend the scope of classical physical explanation, the account of all evidence must be expressed in classical terms. The argument is simply that by the word "experiment" we refer to a situation where we can tell others what we have done and what we have learnt and that, therefore, the account of the experimental arrangement and the results of the observations must be expressed in unambiguous language with suitable application of the terminology of classical physics … [Also we need to recognize] the impossibility of any sharp separation between the behavior of atomic objects and their interaction with the measuring instruments which serve to define the conditions under which the phenomena appear. In fact, the individuality of the typical quantum effects finds its proper expression in the circumstance that any attempt of subdividing the phenomena will demand a change in the experimental arrangement introducing new possibilities of interaction between the objects and measuring instruments which in principle cannot be controlled. Consequently, evidence obtained under different experimental conditions cannot be comprehended with a single picture, but must be regarded as complementary in the sense that only the totality of the phenomena exhausts the possible information about the objects.

Bohr makes three important epistemological points in the above passage. First, he suggests that we cannot separate the observing system or experimental arrangement from the system being observed which, in our case, is the photon whose polarization is being measured. In the act of observation, the observed system interacts with the observing system so that the property observed arises as a consequence of this interaction. The measured property is not a function of the observed system in isolation; in our case the polarization measured is not a property of the photon prior to measurement but arises from its interaction with the detection system. This conception of the origin of the property precludes Einstein's interpretation of the reality principle. What it suggests is that the property is real by the reality principle because it can be predicted in advance; but it is not real in the sense that Einstein construes the principle, because it arises as a result of the particle's interaction with the measuring apparatus and did not exist prior to the measurement context.

Second, there are mutually exclusive experimental arrangements such that the results obtained under one arrangement cannot be combined with those obtained under another incompatible with it. By choosing to measure the polarization along one direction we effectively exclude the possibility of the photon coming to have a well-defined polarization in another direction, since we cannot set up the experimental arrangements to measure polarizations in different directions at the same time. For Bohr, the results obtained under such mutually exclusive experimental arrangements cannot be added together into a single picture as EPR do. We cannot assume that photons have a well-defined polarization in all three directions prior to measurement, since the experimental arrangement to measure the polarization in one direction excludes the possibility of measuring it in other directions.

The third point Bohr makes is that in order to give an unambiguous account of the results of our experiments we have to use the language of classical physics. Thus, even though the phenomena of quantum physics transcend the scope of classical physics, they have to be described—at least the experimental situation and the results of the observation—in classical language. This necessitates complementary descriptions with each having only a limited scope and partiality. Hence Bohr sees both classical and quantum languages as essential elements of the theory, with the classical language being needed to give the results of the observation, and the mathematical language of quantum formalism used to predict the statistical results of the observation. The stress Bohr placed on the need to use classical concepts even after the failure of classical theory is particularly emphasized by Simon Saunders who writes:

> Bohr is crystal-clear. His interpretation is of the phenomena in terms of classical concepts, even though no classical theory can account for such regularities. This was the heart of what was really innovative about Bohr's principle of complementarity: how could one describe regularities classically, when they could not be described by any classical theory? (Saunders 2005: 432)

Bohr's objection to the EPR interpretation of their reality principle can now be formulated as follows. He is not objecting to the principle *per se* but to the way EPR interpret it. This is because the EPR principle of reality itself is formulated ambiguously—it does not tell us whether the property that is predicted in advance, originates in the context of the measurement situation, or preexists this context. In short, the principle allows us to construe two very different kinds of properties as having an element of physical reality—properties that are real because they exist prior to measurement; and properties that are real but get created in the process of measurement. Einstein interprets the real property as preexisting the context of measurement, though this is by no means implied by his principle of reality. By contrast, Bohr takes the property to be real although it originates with the measurement context.

The difference between the two elucidations of the reality principle is of fundamental significance for deciding whether quantum theory is incomplete. It would be incomplete if the complementary measured properties existed prior to the act of measurement. Then it is reasonable to add together into one picture the results obtained under mutually exclusive measurement situations. Since EPR assume the preexistence of the observed properties they are prepared to add them together. However, if Bohr is correct, and these properties arise in the context of measurement, then we cannot legitimately add them together, even though they can each be predicted in advance. To do so is to assume that they exist simultaneously.

Bohr's interpretation of Einstein's reality principle also explains why he considers that we cannot separate the observing and observed systems when we attribute a property to the observed system; and why he argues that the language of classical physics has to be used to describe the results of experiments "though the phenomena of quantum physics transcend the scope of classical physics." Since the measured property is the outcome of the interaction of the measured and measuring system, it cannot be attributed to the measured system retroactively as a property that it possessed prior to its entrance into the measurement context. It is a property that originates in the measurement context and then gets represented in the terms of the language of classical physics.

Nevertheless, many philosophers think that Bohr has confused epistemological and ontological claims. At the heart of such reservations is their belief that Bohr is arguing that we should not simultaneously attribute complementary properties to micro-objects, such as a well-defined position and a well-defined momentum, simply because we cannot measure them simultaneously. They conclude that Bohr is committing the logical fallacy of inferring that properties that cannot be simultaneously measured or observed cannot simultaneously exist. This would be the case only if we assume that what is real is what is observable—precisely the view closely linked to the philosophy of positivism.

These reservations also get strengthened by Bohr's tendency to buttress support for his position by drawing parallels between his construal of quantum properties and Einstein's earlier positivist orientation to space and time when he developed the Special Theory of Relativity. Bohr's appeal to positivist notions is not surprising—at the time he proposed the complementarity viewpoint the positivism of the Vienna Circle dominated the philosophy of science. Consequently many scientists and philosophers linked complementarity to philosophical positivism, including Einstein. According to Folse "this fact may well have strengthened Einstein's conviction that Bohr was approaching the quantum problem from the point of view that all a theory need do is predict correctly the observable phenomena" (Folse 1985: 145–146).[12]

One important additional factor that may have led Bohr to incline toward philosophical positivism was his inability to give a classical realist account of quantum properties. Such classical realism assumed, as EPR did, that the properties measured preexisted in the observed entity prior to measurement. Even today the complementarity standpoint is often seen as embracing an antirealist position—one that puts it securely in the postmodern camp. However, we find that there are good grounds for assuming that complementarity epistemology can be sustained within a realist conception of scientific knowledge, but one which is different from classical realism and separates it from epistemological positivism.

In fact, most scientists who adopt the complementarity framework are realist, rather than positivist, in their construal of the relationship of scientific theories to the world and the properties they measure. Moreover, although the influence of positivism has declined over the last six decades to such an extent that there remain few philosophers or scientists who take it seriously, it has not led to any correspondingly significant decline in the commitment of much of the scientific community to the complementarity interpretation of quantum theory. Most scientists continue to think complementarity works even long after they have turned away from positivism. Hence, there is no reason to think that complementarity necessarily requires us to espouse epistemological positivism.

Furthermore, treating complementarity as a version of positivism would obscure the novelty of its epistemological orientation. It has to be kept in mind that positivism historically predated quantum theory, and nothing in its doctrines makes it more consonant with quantum, rather than classical, physics. Indeed its greatest proponents—Ernst Mach and Richard Avenarius—lived in the nineteenth century long before there was any awareness of the problems raised by quantum theory.[13] If complementarity is designed to bring out the contrast between the way physical properties are understood in classical and quantum physics, then linking it to positivism would only mask, rather than clarify, the revolutionary implications of both quantum theory and its complementarity interpretation.

Moreover, it is doubtful that a positivist account of Bohr's position can be sustained if we consider the explanatory role he assigns to complementarity. Folse makes this point when he notes that belief in the real existence of micro-objects is a prerequisite for Bohr's view of complementarity:

> The phenomenalist interpretation of complementarity misses the point that what is most revolutionary about this framework is the suggestion that in the combination of complementary descriptions of phenomenal objects, we convey information about an independent physical reality. The very fact that it is possible, within the context of quantum theory, to combine these descriptions in a complementary fashion tells us something about an object which is the grounds of these phenomena. (Folse 1985: 243)

Furthermore, Bohr also assumes that the need for complementary descriptions arises because the unobservable micro-object interacts with the observing system to acquire the observed property. This cannot be acceptable to positivists. Positivists cannot allow talk of unobservable entities except as convenient fictions. Hence, they cannot go along with Bohr's account of how complementary descriptions arise, because one cannot speak of the interaction of what they see as a fictional atomic entity with a real measuring apparatus. But interaction between object and measuring instrument is precisely what Bohr invokes to explain the need for the complementarity framework. He writes:

> The apparent contradiction (in complementary descriptions) … discloses only an essential inadequacy of the customary viewpoint of natural philosophy for a rational account of physical phenomena of the type with which we are concerned in quantum mechanics. Indeed the finite interaction between object and measuring agencies conditioned by the very existence of the quantum of action entails—because of the impossibility of controlling the reaction of the object on the measuring instruments if these are to serve their purpose—the necessity of a final renunciation of the classical

> ideal of causality and a radical revision of our attitude towards the problem of physical reality. (Bohr 1935: 696–697)

It is clear that Bohr presupposes the existence of the microsystem in order to speak of its interaction with the measuring agency—a presupposition that would not be sustainable on a strictly positivist stance.[14]

Folse also argues that Bohr's views can only achieve coherence if his epistemological position is undergirded by a new ontological orientation. He writes:

> Because Bohr redefines 'objectivity' in the manner he proposes, in moving from describing systems in classical mechanical states to describing them in quantum mechanical states the system which is *in* such a state can no longer be regarded as a '*substance possessing properties*' precisely because what we predicate of the system in such a state is not, in general, regarded as the 'properties' possessed by a substance. Now in the quantum description, the 'system' which is *in* a quantum mechanical state must be reconceived as an *interaction* which has a feature of 'wholeness' or 'individuality' that implies that the distinction between 'object system' and 'observing system' necessary for making a description of that interaction 'unambiguous' is *relative* to the context of the description. … Bohr leaves us in the dark ontologically, when it comes to the sort of non-ordinary 'reality' we *are* to ascribe to the objects of atomic physics. This reticence in making ontological claims left this dimension of complementarity undeveloped. So from here on we must extrapolate from—or perhaps reconstruct—what Bohr has told us about his 'new viewpoint.' Bohr developed an *epistemological* lesson because his argument terminated in a conclusion about how we gain scientific *knowledge* about physical systems at the atomic level. But epistemological claims about what kind of knowledge we can or cannot have about such systems have *ontological* implications about what such systems which are in quantum mechanical states must *be*. Until such implications are understood, the 'foundations' of the quantum theoretical description of matter will remain mysterious. (Folse 2001: 11)

Folse then proceeds to give an ontological account of the object of quantum mechanical description as an interaction rather than a substance possessing properties:

> [In classical physics] the predicative assertions we make about *them* [objects] can be treated as attributing properties to substances … When we treat statements about the things described by quantum mechanics in the same way,

> because of the quantum postulate, we are led to paradoxical conclusions. Therefore, the object of our description, the 'system' which is 'in a quantum mechanical state,' cannot be a substance possessing properties, but must be regarded as whole phenomena which theory allows us to interpret as an *interaction* between measuring system and the object—the 'atomic system'—of which a measurement outcome is predicated. (Folse 2001: 12)

However, giving an ontological interpretation in terms of interactions, in the way Folse does, is also problematic. It does not explain why the property, say momentum or spin, acquired by an atomic object is retained by it even after it leaves the measuring system. We now proceed to demonstrate that a better ontological account can be achieved by adopting the hypothesis that quantum properties are grown. Although it might appear strange to speak of grown quantum properties, since grown properties are normally associated with biological systems, we find that it has many advantages. First it would preclude the need to embrace a positivist epistemological defense of the complementarity principle quite incompatible with Bohr's recognition of a real microsystem that interacts with a measuring apparatus. Second it would also satisfy Einstein's reality principle without altering its original formulation, but pave the way for a different interpretation of it. Third it would show how Bohr's characteristic epistemological views linked with the complementarity perspective naturally follow from the hypothesis that quantum properties are grown.

Let us now see how these advantages can be demonstrated by treating the polarization of the photon as a property grown in the polarization measuring apparatus in the Aspect et al. experiment. Treating polarization as a grown property explains why we cannot attribute the polarization property measured to the photon even before it enters the apparatus. The measured property grows in the environment of the measuring apparatus as a result of the photon's interaction with it. Consequently Bohr is correct to maintain that the act of observation "disturbs" the observed system and that the measured property is an outcome of this interaction. It also explains why EPR are wrong to assume that their principle of reality shows that the measured property existed prior to measurement simply because we can predict it in advance. What was predicted in advance was nothing more than the property that would grow in a particular measuring environment.

Moreover, assuming quantum properties to be grown explains why the results of measurements from mutually exclusive experimental contexts cannot be added together. Prior to measurement the photon did not have polarization in any direction. Only after its entry into the measuring environment did its polarization grow along the axis defined by the experimental context. Moreover, setting up an arrangement to make its polarization grow along one axis precludes the possibility of setting up an alternative arrangement to make it grow in another direction. Thus the results from such mutually exclusive measurement contexts cannot be combined together since the properties cannot be made to grow together.

The hypothesis that polarization is a grown property also explains why once a particle has been experimentally found to have its polarization in a particular direction it retains this property even after leaving the measuring apparatus. The retention of the property is revealed by the fact that when subsequent attempts are made to measure the same property they yield the same value. If quantum properties are grown properties we would expect this to be the case. The first measuring instrument would be the environment in which the property grew, but the second measuring apparatus merely detects this grown property now carried by the photon. This follows because although grown properties originate in relation to their context, they also subsist outside their originating context and can, therefore, be carried into new contexts.

Finally, seeing quantum properties as grown also allows us to appreciate Bohr's claim that we need to use the language of classical physics in order to give an unambiguous account of the results of experiments. The state of the photon before its polarization is measured—or grown in the process of being measured—is described by the wave function in terms of variables that refer to the polarization it has the potential to grow in any of the three directions. This means that it is not described in terms of properties it actually possesses, but only in terms of properties it could acquire in mutually exclusive experimental contexts. Hence, the equation does not give a concrete and consistent description of the photon prior to measurement, since the physical state of the system is described in terms of variables that refer to the polarization it can come to acquire in different directions, rather than any property of polarization it actually possesses at the time.

The same may be said of properties like the position and momentum of an atomic particle—the properties in terms of which EPR formulated their paradoxical thought experiment. A particle does not have, prior to its entry into the appropriate measuring environment, either a well-defined position or a well-defined momentum. Both position and momentum are grown properties—and the environment in which one kind of property can be grown excludes the environment in which the other can grow. However, the quantum state description of the particle prior to measurement uses the vocabulary of position and momentum combined together—it uses a language formulated in terms of properties it can grow in different mutually exclusive measurement contexts. This vindicates Bohr's claim that in quantum theory we need to use the language of classical physics, such as position and momentum, in order to formulate the theory and give an unambiguous account of the results of experiments, although these properties are not present in the quantum system prior to measurement.

The significance of treating position, momentum, and spin as grown properties is that it accommodates the notion that complementarity involves the features of 'mutual exclusion' and 'joint completion' in descriptions of atomic objects. However, there are some writers, such as Carsten Held, who think that wave-particle complementarity cannot be accommodated into this viewpoint. He writes:

> Consider, for example, Bohr's very accurate description of the two-slit experiment. This experiment can be performed in such a way that always only one particle is contained in the apparatus and hits the screen at a definite position and time. Nevertheless, in the long run, the particles produce the puzzling interference pattern. Now, do the objects in this experiment behave as waves or as particles? Since the single impacts and the interference fringes they form are observed, there is no question here of excluding one classical picture. Some properties of both pictures are somehow 'blended': the determinate position of each particle that hits the screen certainly is a particle property; the distribution of all impacts exhibits interference, thus is a wave property. Hence, wave and particle properties appear in one well-defined experimental arrangement, and there is no question of mutual exclusion of them in the sense of their attribution to exclusive experimental arrangements. (Held 1994: 881–882)

Such a conclusion, however, is disputable. The notions of wave and particle are not mutually exclusive because they cannot be attributed to the quantum object at the same time. They are mutually exclusive because we cannot visualize—put into one picture—their simultaneous existence. We have seen this emphasized by Bohr when he wrote that quantum mechanics requires "*pictures not combinable on the basis of classical physical theories*". And yet it is precisely by combining the mutually exclusive pictures of wave and particle that we achieve joint completion in the description of an atomic object.[15]

Thus treating quantum properties as grown gives a more coherent account of Bohr's epistemological views than the positivist or postmodern positions often attributed to him. It brings together the various interlocking principles of his complementarity viewpoint in a cogent fashion—the impossibility of separating from the measurement context the property observed, the persistence of the observed property beyond this context, the unacceptability of adding into one picture the results obtained from mutually exclusive measurement contexts, and the need to deploy the classical language of physics that refers to properties.

Moreover, it also enables us to understand why Einstein saw quantum theory as incomplete. By assuming that the object possessed the property grown in a measuring environment, even before it entered that environment, simply because we can predict what it would be before measurement, he mistakenly concludes that quantum theory is incomplete. However, this is not a tenable conclusion if what is predicted in advance is nothing more than the property that will grow in the selected measuring situation. Indeed by precluding us from ascribing a grown property to an atomic system, which does not have it until it gets grown in the process of measurement, quantum theory accurately represents reality—it does not, like Einstein, wrongly attribute a grown property to a system even before it has grown.

2.2 Complementarity of Genetic and Environmental Explanations

Clearly treating quantum properties as grown enables us to make sense of Bohr's complementarity perspective and repudiates the charge that his views are incoherently both positivist and realist at the same time. It also supports his claim for the completeness of quantum theory. Moreover, since quantum properties are grown, and thus similar to biological properties, Bohr's view that complementarity can be extended to illuminate biological phenomena seems quite reasonable. We now proceed to show how complementarity extended to biology—what we have termed "biological complementarity"—can illuminate our understanding of processes in developmental biology, evolutionary biology, and ecology.

Let us begin by considering how the epistemology of complementarity can illuminate our understanding of processes in developmental biology. Central to developmental biology is morphogenesis—a process in which cells in a developing embryo multiply by division to produce an organism—the phenotype—from a fertilized seed or egg—the genotype. It is generally assumed that the genetic code of the genotype carries the information for producing the phenotype. However, there are different ways in which the genotype can be seen as carrying such information for the production of the phenotype. First, the genotype can be seen as containing a blueprint in explicit detail for the phenotype, as the design plan of a watch lays out the way parts should be shaped and assembled in its construction. Second, the genotype can be seen as carrying a general program for generating the phenotype which is sensitive to the environmental context in which it grows, like a program of a chess playing computer that adjusts moves in response to moves made by its opponent. Third, the genotype could be taken as conveying a recipe for the production of the phenotype analogous to the way a cookbook carries a recipe for making a cake—by giving instructions for combining ingredients taken from the environment to produce a finished outcome.[16]

All three metaphors—blueprint, program, and recipe—have been deployed by developmental biologists to show how a genotype produces a phenotype or how the genetic code generates an organism. Nevertheless, most biologists today would not assume that the genotype carries information intended for the production of a precisely well-defined phenotype. Many would also argue that it is more illuminating to see the genotype as not carrying specifications for a generally predetermined phenotype, but only instructions for producing new cells in response to the specific environmental context in which such cells grow. This is because changing the environment in which a genotype expresses itself can change the resulting phenotype. The dependence of the properties of the phenotype on the environment makes it difficult to envisage that the genotype carries information for generating a precisely specified phenotype. It is the role of the environment in determining the structure of the phenotype which arises from the genotype, as the role of the experimental context in determining the specific quantum property that arises, which makes the complementarity perspective relevant for illuminating processes in developmental biology.

To appreciate this point let us begin by examining the view that the genotype carries a plan or a blueprint for an organism. Such a view assumes that specific genes carry information for specific parts of an organism. Taking the example of an organism such as a frog, it assumes that there are eye genes for forming its eyes, liver genes for its liver, muscle genes for its tail, and so on. Adopting this view suggests that when a frog embryo develops from a fertilized egg it is being guided by its genetic blueprint, so that the process of development unfolds naturally provided the environment of the growing organism furnishes sufficient energy and nutrition. It implies that different parts of the genetic code carry information about different parts of the organism, and the blueprint controls the division of cells so that the organism specified in the blueprint gets finally assembled. Such a blueprint conception of the genetic code dominated much early thinking in molecular biology, but hardly any biologists currently subscribe to it in its full-fledged form today.

The reason is that there are many problems with the notion of a genetic blueprint. It does not explain why, despite the fact that every cell of a developing embryo carries the same genetic code, embryonic development leads to cell differentiation. For example, in the frog embryo we find differentiated nerve, blood, muscle, bone and other types of cells carrying the same genetic code. But the genetic blueprint model makes it difficult to explain why cells ever come to differentiate and specialize if all of them are responding only to the same code in their nuclei. Of course, such a problem would not arise if the original cell from which they descend through cellular division transmits different parts of its genetic code to the different progeny of cells that are born from it. However, this is not the case because every cell of the organism inherits the whole of the genetic code of the original fertilized cell from which it descends. Given that all cells carry the same code why would one cell, say, become a nerve cell and another a muscle cell?

There is one way of getting around the problem. We can treat the genetic code as a program rather than a blueprint for an organism. The program would then unfold as the embryo of the developing organism grows. Even though each generation of dividing cells in the growing embryo inherits the same DNA code, they grow and multiply in a differentiated fashion depending on the instructions they receive from the unfolding program. A cell which becomes a nerve cell, say, in contrast to another that becomes a muscle cell with different properties, reaches its different destination by being directed along a different developmental path by a different part of the genetic program.

Nevertheless, the notion of a genetic program also has problems. Embryonic tissue cells are, as we have seen, far too sensitive to environmental information to allow for such a conception. Such plasticity of response of a cell to its local environmental information makes it difficult to see the developmental process as the unfolding of a predefined program within the genetic code for forming an organism.[17]

We could, of course, deal with this problem by making the program itself flexible. We could see the program as pursuing its goal of actualizing the organism in a way that is sensitive to the environmental context within which cells multiply and develop. Such flexible genetic programs would be similar to computer programs that are goal-oriented but flexible. Indeed, a genetic code that carries a goal-oriented program with flexible strategies would be more likely to reach its final predetermined goal of the phenotype it specifies even in the face of unanticipated environmental contingencies.

However, all the accounts of the genetic code we have examined so far—the code as a blueprint, or an unfolding program, or a flexible program—assume that the genotype carries information to produce a predefined phenotype. This assumption is questionable because there is considerable evidence to show that very different phenotypes can develop from the same genotype under different environmental conditions—what has been labeled as 'phenotypic plasticity'. It suggests that the genotype cannot be taken as the only carrier of information for producing a phenotype—the assumption that a specific phenotype is the goal of the genotype is untenable.

To illustrate this point consider the way the phenotype of many species of plants alter when they grow in different environments even though their genotype is the same. Verne Grant describes an experiment in which the biologist, Anton Kerner, collected the seeds from plants that were generally found in the lowlands and planted them high in the Alps at 7200 feet. The mature plants were quite different from their parents. Their flowers and leaves were smaller and brighter, their stems were shorter and they bore flowers closer to the ground. Nevertheless, when seeds from these plants raised at Alpine heights were again sown in the valley they resumed their lowland characteristics. There was no evidence of any hereditary modification in form or color. Clearly the wider environment has an important role in shaping the phenotype that develops out of a genotype. (Grant 1963: 129). Grant describes another experiment where the plant *Potentilla glandulosa* was brought up in four different environments: dry and sunny, moist and sunny, dry and shady, moist and shady. They produced four strikingly different phenotypes. The plants grown in moist conditions were more luxuriant but the plants grown in shade were taller and had broader leaves (Grant 1963: 121).

Such examples suggest that the phenotype is the outcome of the interaction of the genotype with its environment. Consequently, the phenotype cannot be treated as fully specified by the genetic code carried by the genotype, but only conditioned in part by it. The code constrains the range of phenotypes that may develop, but which one gets realized also depends on the environmental context within which an organism grows. Since the properties of the phenotype are grown from a genotype in dependence on its environmental habitat and are, therefore, the outcome of the gene–environment interactions, the genetic code cannot be treated as the sole carrier of information for producing the phenotype. Different phenotypes could arise from the same genotype under different environmental conditions.

This raises the following question: What are the instructions in the genetic code designed to do if they are not carrying a blueprint or a program for producing a predefined phenotype? An illuminating answer to this question was offered nearly a century ago by the biologist Paul Weiss through the notion of "position effect", which he deploys to account for cell differentiation during the process of morphogenesis. Weiss writes:

> Let us take a circumscribed body, depending for its maintenance on active exchange with its environment; for instance, an egg in a pond, a cell in a tissue, a human individual in society. Then let the unit multiply into a few more units; they all continue to have a share in the common interface of exchange and communication with the medium. But let the number of units keep on increasing, whether by subdivision or accretion, and all of a sudden a critical stage arises at which some of the units find themselves abruptly crowded inward, cut off completely from direct contact with their former vital environment by an outer layer of their fellows. The latter thereby acquire positions not only *geometrically intermediary, but functionally mediatory*, between the ambient medium and the now inner units. From then on, "inner" and "outer" units are no longer alike. A monotonic group of equals has become dichotomized into unequal sets … The train of events to follow such a "differentiation" of a radially symmetrical core-crust dichotomy is easy to envisage. Interactions between the "outer" members and their newly established "inner" neighbors would expose to another set of new conditions any fresh units arising subsequently in the intermediate zone between them, and hence call forth in them a third type of reaction. Moreover, polarized influences from the outside would impose an axiate pattern upon the group. Thus would ensue a train or sequelae of ever-mounting, self-ordering complexity. In all these steps, the fate of a

> given unit would be determined by its response to the specific conditions prevailing at the site in which it has come to lie, those conditions varying locally as functions of the total configuration of the system—its "field pattern", for short. This principle—long recognized empirically as a basic criterion of systems but not always fully appreciated in its implications—is commonly referred to as "position effect". (Weiss 1973: 31–32)

In this passage Weiss gives an account of self-regulation and self-ordering by appealing to the way units differentiate by virtue of their "geometrically intermediary, but functionally mediatory" role. He refers to the factor that produces this outcome as "the position effect." In any growing process as each unit multiplies—whether an egg in a pond, a cell or a human individual—the position effect makes them differentiate by acquiring properties that are conditioned by their local environmental contexts. Hence, even though they may have begun as a "monotonic group of equals" they differentiate in response to varying environmental circumstances that assigns them different functional roles.

Lewis Wolpert deploys a more recent but similar account of cell reproduction to explain pattern formation in biological development.[18] According to Wolpert this process involves two steps: the cells first acquire environmental information and then interpret such information to express their genetic program. It is significant that although Wolpert appeals to the notion of a genetic program, for him the genetic code is not an unfolding or flexible program that implicitly carries an organic form. Wolpert's program is radically different—it contains no information about organic form. It merely tells a cell how to respond to information supplied by its local environment. Whatever organic form results, it is merely the outcome of the sum-total of such local responses, which could vary quite radically depending on environmental conditions.

Nevertheless, it is important to note that Wolpert's model assumes that environmental information is conveyed by a more complex process than the influence of neighboring cells envisaged by Weiss. Wolpert sees the environment as a whole formed by the cells already produced, as creating a gradient field of concentration of a substance—what he terms a *morphogen*—that influences the way future cells differentiate as they read environmental information through their genetic program.

Wolpert's model suggests that the genetic code does not carry instructions to produce a phenotype, but only guidelines for generating new cells in response to environmental factors. The genetic code tells a cell how to reproduce in its local context, a context shaped by an array of influences from diverse processes, but does not determine any specific phenotype. Whatever final phenotype arises would only be the incidental outcome of numerous local processes, where each process is the result of cells responding to environment-sensitive instructions in their genetic code. Consequently a large number of different phenotypes could arise from the same genotype as a result of growth taking place in different environments. Thus, although the genotype does place limits on the potential phenotypes that can be generated from it, it cannot be taken to carry instructions for any specific phenotype.[19]

In everyday experience we do not see a multiplicity of phenotypes arising from a single genotype. This is because the environment in which organisms develop is often kept fairly constant under natural conditions. The genes of animals develop in the womb of animals; the genes of plants in the controlled environment of seeds. Thus the environment for the expression of the genetic code is held fairly constant in all of these cases. Hence, the phenotypic outcome is also fairly well defined. Nevertheless, with different habitats even the same genotype can lead to different phenotypes. Thus, without making arbitrary assumptions concerning what constitutes the normal phenotype carried by the genotype, so that other phenotypic outcomes are seen as deformations of this normal phenotype, we have to consider all phenotypic outcomes under diverse environmental conditions as products carried by the same genetic code. To presume some phenotype as 'normal', and to be the predefined goal of the genetic code, and others as 'deformations' of it would be arbitrary.

The one-to-many relationship between a genotype and the phenotypes that can arise from it enables us to understand why the biological complementarity of the mechanical or molecular and functional or environmental viewpoints recommended by Bohr can illuminate developmental biology. To illustrate this point consider the properties of the plant *Potentilla glandulosa* brought up in four different environments: dry and sunny, damp and sunny, dry and shady, moist and shady. We have seen that it leads to four strikingly different phenotypes from the same genotype. Can each of these phenotypes be explained by adopting merely a molecular viewpoint that sees them as outcomes of instructions solely carried by the genotype? No, because their distinguishing properties are not only a function of the genotype, but also the different environments in which they grew. Neither can we treat any of them as the outcome of the environment alone since they are also an expression of instructions in the genotype. Hence, the distinguishing properties of the different phenotypes have to be explained by combining a molecular or genetic explanation with a functional or environmental point of view that allows us to see how each phenotype adapts the plant to different environmental conditions.

There may be a temptation to argue that the functional explanation is merely a shorthand way of dealing with molecular processes in the environment interacting with the developing organism that are simply too complex to unravel individually. This assumes that, in principle, the molecular point of view is sufficient, but in practice, we might turn to a functional argument simply because the molecular processes are too complex to work through individually. Such an argument would be credible within classical physics, but quantum theory subverts this assumption. All molecular processes ultimately involve the interactions of atomic particles. We have seen that the properties of such particles grow in the measuring context in which they are identified. If the property of even a single atom grows in the environmental context in which it gets measured, so that it becomes impossible to predict the property independent of its generating context, then surely we must assume that the properties of a phenotype cannot be predicted from a molecular point of view. The functional or environmental explanation must be a necessary complement to the molecular explanation.

2.3 The Cumulative Genotype and Disjunctive Phenotype: Two Patterns of Evolution

Ignoring the environmental standpoint in shaping phenotype properties, and assuming that we only need a molecular genetic approach, can also deform our understanding of biological evolutionary processes. This can be illustrated by looking at the problems faced by the biologist Richard Dawkins when he endeavors to give a purely mechanical, molecular, account of biological evolution in his work *The Blind Watchmaker: Why the Evidence of Evolution Reveals a Universe without Design*. Dawkins has since modified the radically reductionist vision he assumed nearly three decades ago but the position he adopts continues to generate controversy even today and merits close investigation.[20]

Consider Dawkins account in this study of how a genotype develops into a phenotype through the processes of morphogenesis. He argues that the information for generating the organism cannot be seen as contained in the genetic DNA code as a blueprint. Instead he maintains that the DNA code should be seen as carrying a recipe:

> Now, we don't yet understand everything, or even most things, about how animals develop from fertilized eggs. Nevertheless, the indications are very strong that *the genes are much more like a recipe than like a blueprint*. Indeed, the recipe analogy is really rather a good one, while the blueprint analogy, although it is often unthinkingly used in elementary textbooks, especially recent ones, is wrong in almost every particular. Embryonic development is a process. It is an orderly sequence of events, like the procedure for making a cake, except that there are millions more steps in the process and different steps are going on simultaneously in many different parts of the 'dish'. Most of the steps involve cell multiplication, generating prodigious number of cells some of which die, others of which join up with each other to form organs, tissues and other many-celled structures. ... There is no simple one-to-one mapping, then, between genes and bits of body, any more than there is a mapping between words of recipe and crumbs of cake. The genes, taken together, can be seen as a set of instructions for carrying out a process, just as the words of a recipe, taken together, are a set of instructions for carrying out a process. (Dawkins 1986: 295–296)

Dawkins also uses his recipe view of the genetic code to explain how biological evolution occurs. He makes his case by giving a computer model that simulates biological evolution by using recursive rules to generate artificial life forms, which he terms 'biomorphs'. (Dawkins 1986: 55) Dawkins' biomorphs are generated by a tree growing procedure that begins with a single line. It then branches into two lines; the branches then split into two sub-branches; and the sub-branches into two sub-sub-branches and so on growing the way a tree grows. The final form results after a predetermined number of recursions. Dawkins defines the depth of recursion needed to generate a biomorph by the number of sub-sub-... branches that are used. He argues that the process through which biomorphs grows is analogous to morphogenetic processes in embryonic development:

> Recursive branching is also a good metaphor for the embryonic development of plants and animals generally. I don't mean that animal embryos look like branching trees. They don't. But all embryos grow by cell division. Cells always split into two daughter cells. And genes always exert their final effects on bodies by means of *local* influences on cells, and on the two-way branching patterns of cell division. An animal's genes are never a grand design, a blueprint for the whole body. The genes, as we shall see, are more like a recipe than like a blueprint: and a recipe, moreover, that is obeyed *not* by the developing embryo as a whole, but by each cell or each local cluster of dividing cells. I am not denying that the embryo, and later the adult, *has*

> a large-scale form. But this large-scale form *emerges* because of lots of little local cellular effects all over the developing body, and these local effects consist primarily of two-way branchings, in the form of two-way cell splittings. It is by influencing these local events that genes ultimately erect influences on the adult body. (Dawkins 1986: 51–53)

The recursive branching rules Dawkins uses to specify his biomorphs determine various traits such as the angle of branching, the length of a branch relative to its predecessor, the depth or number of recursions before the figure is brought to a stop and so on. Dawkins considers these set of rules to be 'genes' for the biomorph since they generate it in the way a genetic code generates a biological organism by rules it carries.

To introduce the idea of evolution into his model, Dawkins examines how biomorphs change when one alters their generating rules, or 'genes', in small discrete steps, one rule at a time. This is analogous to introducing small variations in the genetic code of a genotype. For example, the depth of recursion rule could be altered by increasing it by +1 or reducing it by −1, keeping the other rules unchanged. This would alter the form of the biomorph by either increasing or decreasing the branchings by one step. Similarly one could alter the rule for the angle, the length and so on. Dawkins has nine 'genes' or rules for his model that he can alter to shape the figure that forms. What is important is that Dawkins generates each biomorph, which he calls a child, by altering only one gene in the complex of genes that generated its parent, and confining this alteration to a small prespecified +1 or −1 increment in values. He writes:

> The shape of each child is not derived directly from the shape of the parent. Each child gets its shape from the values of its own nine genes (influencing angles, distances, and so on). And each child gets its nine genes from its parent's nine genes. This is just what happens in real life. Bodies don't get passed down the generations; genes do. Genes influence embryonic development of the body in which they are sitting. (Dawkins 1986: 55–56)

Since in each generation there are nine genes, only one of which can be altered by either a +1 or −1 direction in every generation, there are 18 potential biomorphs that could descend from one parent biomorph. For each generation Dawkins selected one of these 18 forms—the one he considered to be the most aesthetically appealing—to be the parent of the next generation. He argues that this artificial selection is analogous to the way natural selection determines which biological phenotypes will survive and reproduce.

Moreover, as time went on, and Dawkins gradually increased the number of recursions to generate the biomorphs, the resulting forms became more complex and interesting. The first generation of biomorph was merely a small straight line; the second had the shape of the letter Y; by the seventh generation the biomorph assumed the form of a catapult; by the tenth generation it looked like a section through a flower; and so on until the in the 29th generation the biomorph resembled an insect of sorts.

After describing the outcomes of his artificial selection Dawkins writes:

> All these children are mutant children of the same parent, differing from their parent with respect to one gene each. This very high mutation rate is a distinctly unbiological feature of the computer model. In real life, the probability that a gene will mutate is often less than one in a million. The reason for building a high mutation rate into the model is that the whole performance on the computer screen is for the benefit of human eyes, and humans haven't the patience to wait a million generations for a mutation! (Dawkins 1986: 56–57)

Dawkins then goes on to emphasize that his model illustrates the cumulative and incremental nature of the process of biological evolution:

> Notice how each generation is just a little different from its parents and from its sisters. Since each is a little different from its parents, it is only to be expected that each will be slightly *more* different from its grandparents (and its grandchildren), and even more different still from its great grandparents (and great grandchildren). This is what *cumulative* evolution is all about. (Dawkins 1986: 59) [Dawkins' emphasis]

Dawkins designed his biomorph model to mimic cumulative evolution in biology by showing how small cumulative changes in the rules generating the biomorphs will also lead to small cumulative changes in biomorph forms. But he notes in surprise that this was not what he actually observed since he found that his model was "more like a pedigree of species than a pedigree of individuals". The changes across each generation were often sharper than his cumulative model predicted. He proceeds to dismiss this anomaly to his position by attributing it to the artificially high mutation rate he has imposed on his model (Dawkins 1986: 43).

However, Dawkins' explanation for what he observed of biomorph evolution is wrong, although the disjunctive changes he observes in the pattern has significant implications for our understanding of biological evolution. The actual reason why his forms appear more like a pedigree of species than individuals is that the rules Dawkins uses to generate biomorphs are not like recipe rules used to produce a cake. A recipe gives a set of rules for a sequence of steps to produce something, so that different rules apply at different stages of the production process. The recipe rules for making a cake would change at different stages of its production. But the rules Dawkins uses for generating his biomorphs are recursive rules that are repeatedly applied as the biomorph develops from parent to child.

Hence, the rules Dawkins adopts to generate his biomorphs are more like the rules that generate complex fractal forms than the rules we use to make cakes. It is widely recognized that we can produce very complex and rich fractal forms by using very simple rules which get recursively applied. Moreover, small changes in such recursive rules can lead to dramatic changes in the resulting forms. The crucial difference between recursive and recipe rules is that recipe rules are not applied repeatedly at every step in the production of a cake, but change at different stages in the cake-making process.

It is precisely the difference between recursive and recipe rules that explains the large changes in his biomorphs that Dawkins observes occasionally across single generations, despite the small changes in the rules he makes. It surprises him because we would not expect this on a recipe view—changing the quantity of the ingredients for a cake marginally would not dramatically alter the overall quality of the cake. By contrast, once we recognize biomorph generating rules as really recursive rules we can understand why even small alterations in them lead to large alterations of form. Although a small change in a recipe rule always involves a small change in the outcome, a small change in a recursive rule can get amplified, through the successive steps in which it is iteratively applied, to produce a large change in the final outcome. Dawkins' biomorph does not grow like a cake but like a fractal form.[21]

Indeed, had Dawkins actually changed the biomorph generating rules at each stage of the biomorph's development step by step, as in a recipe, he would have found that the resulting forms would also have evolved more gradually in conformity with the way he expects evolution to proceed. Hence, even if Dawkins is correct to see genotypic evolution as proceeding in small cumulative steps, he is wrong to use this as a reason for supposing phenotypic evolution should also exhibit the same pattern. Since the instructions in the genotype get recursively applied we should expect even small variations in the genotype to sometimes lead to large variations in the resulting phenotype. Hence, biological variation can appear either cumulative or punctuated when seen from the perspectives of the changes in the genotype, and its associated phenotype, over long historical periods.

This suggests that organic evolution can exhibit two perspectives depending on whether we view it in terms of the changes in the genotype or the phenotype, and that the pattern of genotype evolution need not conform to the pattern of phenotype evolution. Even though genotypic evolution generally occurs through small, incremental, isolated steps it may trigger a pattern of phenotypic evolution that involves large, punctuated, and integrated changes. The only constraint is that any variation produced in the phenotype has to be adapted to the environment in which it lives and reproduces—otherwise it will get eliminated by natural selection.

Hence, when we speak of evolution as proceeding through random variations subject to natural selection it is imperative to keep in mind that there are two kinds of organic variations—genotypic and phenotypic—upon which natural selection operates. We can afford to ignore this distinction only if we assume, as Dawkins does, that small genotypic variations invariably produce small phenotypic variations. Then we can speak of organic evolution as having a single structural pattern where gradual genotypic evolution necessarily implies gradual phenotypic variation. However, this cannot always be expected if the phenotype is generated by application of recursive rules carried by the genotype. Then it is possible for genotypic evolution to proceed in small, incremental, isolated steps while at the same time phenotypic evolution exhibits a pattern of large, disjunctive, and integrated changes.

Let us look at three classes of biological phenomena in nature that illustrate the importance of taking into account the distinction between genotypic and phenotypic variations—punctuated equilibria, organs of extreme perfection, and overshoot phenomena. The notion of punctuated equilibrium suggests that the evolution of species occurs not gradually but in a jerky fashion so that periods of slow cumulative change alternate with sudden rapid transformations. The phenomena of organs of extreme perfection, such as the eye, appear to involve simultaneous coordinated changes in many different organ systems to evolve successfully. Finally, we have overshoot phenomena, such as the peacock's elaborate tail, in which the evolved trait seems to reduce the adaptability of the organism to its environment. These three classes of phenomena are taken by Dawkins as particularly problematic for his recipe view of the genetic code and receive special attention from him. In order to deal with them Dawkins invokes additional auxiliary hypotheses, with specifically different assumptions for dealing with each type of phenomena, to show they conform to Darwin's theory of evolution. However, we will find that adopting a complementarity perspective allows us to deal with them without such auxiliary hypotheses and without violating the fundamental notions of variation and natural selection integral to Darwin's theory.

Let us first begin by looking at punctuated equilibrium theory. Punctuationalism has had a fluctuating history in biology and seems to be strongly supported by fossil evidence. When we arrange all known fossils in chronological order they reveal a jerky sequence in which there are sharp gaps between neighboring species. This appears to conflict with Dawkins' notion that evolution proceeds by means of small random changes in the genetic code of organisms resulting in a gradual and continuous sequence of morphological changes in phenotypes. Dawkins resolves the problem by arguing that fossilization occurs only under special circumstances so that many intermediate steps in organic evolution have not left any traces in the fossil record. Hence, even if the actual process of phenotypic evolution had been continuous, it would appear to be jerky if we depend only on the fossil evidence. Moreover, even those organisms that get fossilized do not always get discovered—it is only when we happen to chance on the fossil record that we can collect the evidence it provides. Dawkins compares looking at fossil evidence as being similar to watching a movie in which most of the frames have gone missing, except for a few isolated frames that happen by chance to have been saved and discovered.

Indeed Darwin himself was disturbed by observed data that seemed to suggest sudden large-scale changes of form because they threatened his belief that evolution of phenotypes should be gradual and mediated by small random variations. Hence, he was not only concerned about the absence of intermediate species in many cases within the geological record, but also dismayed that changes of organic form sometimes appeared to occur in disjunctive integrated jumps. However, he held on to the principle of gradual evolution of phenotypes because he thought that anyone admitting abrupt and integrated changes in their evolution "calls in the agency of a miracle."[22] This led him to explain the jerky fossil record by appealing to gaps in this record. This is precisely how Dawkins also attempts to save his recipe view of the genetic code.

However, if we take the genetic code to carry rules, which get recursively applied in the process of cell multiplication that generates the phenotype, then there is nothing miraculous in sudden and macroscopic changes in phenotypes corresponding to microscopic variations in the genotype. This follows from the fact that even minor alterations in genetic rules can get amplified through the morphogenetic process when they get recursively applied to form the phenotype. What would be miraculous would be sudden large integrated variations in the genotype, not such macro-variations in the phenotype. Darwin's theory appears threatened only when we conflate, and fail to distinguish, these two different kinds of variations involved in biological evolution—micro-variations of the genotype and correlated macro-variations in the phenotype.[23]

To make the assumption that a micro-variation in the genotype always leads to a micro-variation in the phenotype is to ignore the fact that between the genotypic variation that affect the genetic rules and the phenotypic variation that affects the organic structure, there intervenes a process of embryogenesis in which the genotype produces the phenotype in interaction with the environment. Embryogenesis can amplify dramatically a micro-variation in the genotype into a macro-variation in the phenotype. Consequently it is possible for macro-variations in the phenotype to evolve in parallel with genotypic micro-variations. The jerkiness in the fossil record of phenotypes would then not be the outcome of an incomplete record of the pattern of evolutionary change, but a faithful record of this pattern.

Darwin himself gives direct evidence for such a radical phenotype change without intermediate forms, which came to be preserved by artificial selection. He writes:

> In some few instances new breeds have suddenly originated; thus, in 1791, a ram-lamb was born in Massachusetts, having short crooked legs and a long back, like a turnspit-dog. From this one lamb the otter or ancon semi-monstrous breed was raised; as these sheep could not leap over the fences, it was thought that they would be valuable; but they have been supplanted by merinos, and thus exterminated. These sheep are remarkable from transmitting their character so truly that Colonel Humphreys never heard of "but one questionable case" of an ancon ram and ewe not producing ancon offspring. When they are crossed with other breeds the offspring, with rare exceptions, instead of being intermediate in character, perfectly resemble either parent; and this has occurred even in the case of twins. Lastly, "the ancons have been observed to keep together, separating themselves from the rest of the flock when put into enclosures with other sheep." (Darwin 1868: 126)

It is significant that Darwin noted ancons to not only appear different from normal sheep but also to have a tendency to separate themselves from normal sheep. This makes it possible to envisage that, left by themselves in the wild, ancons would spontaneously separate themselves from other sheep and take a different evolutionary path until they eventually develop into a separate species.

It is also significant that punctuated variation precedes separation from a mother species in the case of ancons above. By contrast, contemporary punctuationalists explain the punctuated fossil evidence by seeing separation as preceding the emergence of punctuated differences. They envisage a daughter-species separating from an ancestral species as a result of geographical isolation—e.g. a daughter-species being, for some reason, cut off by mountains, rivers, and so on from their ancestors who may continue little changed. This separation prevents interbreeding so that after a long period of gradual evolution they would become a different species unable to interbreed with other descendants of their common ancestor even if they returned to their original habitats. Although the change occurred gradually it would leave a jerky record in fossil evidence simply because the evidence for continuity may be difficult to discover without coming upon—generally by accident—the isolated geographical area over which the evolutionary change developed.

This is how Stephen Jay Gould explains the punctuated fossil record in his study *The Panda's Thumb*:

> A new species can arise when a small segment of the ancestral population is isolated at the periphery of the ancestral range. Large, stable central populations exert a strong homogenizing influence. New and favorable mutations are diluted by the sheer bulk of the population through which they must spread. They may build slowly in frequency, but changing environments usually cancel their selective value long before they reach fixation. Thus, phyletic transformation in large populations should be very rare—as the fossil record proclaims.
>
> But small, peripherally isolated groups are cut off from their parental stock. They live as tiny populations in geographic corners of the ancestral range. Selective pressures are usually intense because peripheries mark the edge of ecological tolerance for ancestral forms. Favorable variations spread quickly. Small, peripheral isolates are a laboratory of evolutionary change.

> What should the fossil record include if most evolution occurs by speciation in peripheral isolates? Species should be static through their range because our fossils are the remains of large central populations. In any local area inhabited by ancestors, a descendant species should appear suddenly by migration from the peripheral region in which it evolved. In the peripheral region itself, we might find direct evidence of speciation, but such good fortune would be *rare* indeed because the event occurs so rapidly in such a small population. Thus, the fossil record is a faithful rendering of what evolutionary theory predicts, not a pitiful vestige of a once bountiful tale. (Gould 1990: 183–184)

However, once we recognize that even a single genotypic micro-variation can produce a macroscopic phenotypic variation it is possible to see how swift speciation can occur even without geographical isolation. In this respect Darwin's ancon example is illuminating. It illustrates how a radical macro-variation of the phenotype can arise from a random micro-mutation in a genotype. In the case of the ancon sheep the separation was effected by human beings who isolated the mutated sheep from the others and deliberately selected it for reproduction. However, the same process could occur naturally if the mutation enables the organism to exploit a new geographical niche not accessible to other members of its species. As more members with the mutated trait exploit the niche the new variation will become increasingly isolated from the parent group. The process is similar to what Gould describes, except that *the punctuated variation precedes the geographical isolation, and not vice versa*. This hypothesis suggests that geographical isolation could follow punctuated changes, and that a punctuated fossil record may faithfully render the actual process of change.[24]

The possibility of the occurrence of phenotypic macro-variations correlated with genotypic micro-variations is by no means purely speculative. Apart from ancon sheep there are many examples of macro-variations in the phenotype arising from a micro-mutation in the genotype. Homeotic mutations illustrate how even mutations in a single gene can lead to large changes in organic form.[25] One example of homeotic mutations involves pea-plants which normally have leaflets near the base and tendrils at the tips. It has been observed that one mutation in a single gene in these plants causes all their leaflets to be replaced by tendrils; another mutation in a different gene reverses the effect and causes the tendrils to be replaced by leaflets.[26] Another example of homoeotic mutations involves the ubiquitous fruit-fly—one mutation in one gene in the fruit fly replaces all its antennae by legs giving rise to so-called *antennapedia* mutants; another mutation in a single gene cause a pair of legs to be replaced by antennae. A third type of mutation causes flies to develop four-wings instead of two. All of these examples suggest that disjunctive large-scale variations in the phenotype can arise from small mutations in the genotype. This suggests that there is no need to invoke additional auxiliary assumptions, such as gaps in the fossil record, to explain the phenomena of punctuated changes in the evolution of species, since punctuated phenotypic changes can be the result of small cumulative genotypic changes. Seeing organic evolution as gradual and cumulative from the molecular genotypic perspective, and sudden and disjunctive from the holistic phenotypic perspective, can be understood as different perspectives on the pattern of organic evolution.

However, it is important to note that the genotypic and phenotypic perspectives are not mutually exclusive but jointly necessary perspectives—they are mutually inclusive though jointly necessary to completely understand the pattern of organic evolution. What exhibits complementarity is the functional explanation of the phenotype properties in terms of adaptation to the environment and their molecular explanation in terms of its genetic code. Many phenotypic traits can only be explained by combining the genetic and environmental influences, which are often inextricably entangled. This is precisely why the gene versus environment debates in morphogenesis have generated so much controversy reminiscent of the wave and particle debates in physics. This is precisely why seeing them as complementary, in the way Bohr recommended, can be illuminating.

Let us now turn to a second class of apparent problems for the Darwinian account of evolution that Dawkins addresses. Many observations suggest that phenotypes evolve through processes that bring about changes of form in two different ways—variations in the overall shape or size of organisms as a whole, and variations involving integrated changes in the diverse organ systems in organisms. Both create problems for the notion that organic evolution of form should involve small cumulative steps of separate changes across different parts of the organism, rather than in coordinated transformations over the whole organism.

There are many examples in nature of the first kind of variation. The body of the sunfish has a form that is a topological transformation of the porcupine fish. By drawing the porcupine fish on a rubber sheet we can change it into the form of a sunfish by stretching the sheet appropriately. Similarly, the shape of the baboon skull can be related to that of a chimpanzee or a human by other topological transformations. Such transformations of organic forms have been studied systematically by D'Arcy Thompson who writes:

> We know beforehand that the main difference between the human and the Simian types depends upon the enlargement or expansion of the brain and braincase in man, and the relative diminution or enfeeblement of his jaws ... [W]e are not shewn by the ordinary methods of comparison, how far these various changes form part of one harmonious and congruent transformation, or whether we are able to look, for instance, upon the changes undergone by the frontal, the occipital, the maxilliary and the mandibular regions as a congeries of separate modifications or independent variables ... [But once we recognize the others are] a simple 'projection' of our human skull ... it becomes at once manifest that the modifications of jaws, braincase, and regions between, are all portions of one continuous and integral process. (Thompson 1942: 1082–4)[27]

It is sometimes argued that such global changes of form cannot be explained by natural selection operating on random variations. It has also been suggested that the existence of such integrated transformations of form across species reveals that evolution is guided by the existence of hitherto undiscovered "laws of form."[28]

However, by recognizing that even small isolated changes in the genotype can bring about large integrated changes in the phenotype, we can give a Darwinian account of such changes of shape without appealing to new laws of form that direct biological evolution. Indeed small isolated mutations in the genotype can bring about large coordinated changes in the global form of an organism, because the effects of these mutations can become amplified in an integrated fashion through the morphogenetic process of cell division and multiplication that leads to the formation of the organism. Since such a genetic mutation affects the organism as a whole, it cannot be surprising that successive stages of the evolution of forms are often topological transformations of earlier stages.

The same argument may be extended to explain changes across different organ systems that have to occur together as an integrated complex in the organism to confer an evolutionary advantage. Darwin pointed to such examples when he raised it as the problem of the evolution of the eye.[29] In his book *The Neck of Giraffe*, Francis Hitching formulates the problem as follows:

> The eye either functions as a whole, or not at all. So how did it come to evolve by slow, steady, infinitesimally small Darwinian improvements? Is it really plausible that thousands upon thousands of lucky chance mutations happen coincidentally so that the lens and the retina, which cannot work

> without each other, evolved in synchrony? What survival value can there be in an eye that doesn't see? (Hitching 1982: 85, quoted in Dawkins 1986: 80)

Indeed Darwin wrote that "The eye to this day gives me a cold shudder, but when I think of the fine known gradations, my reason tells me I ought to conquer the cold shudder."[30] He also saw the eye as but one example of what he termed "organs of extreme perfection" which he saw as problematic for the notion of evolution as a process of small and isolated cumulative changes in an organism.

Dawkins attempts to explain the evolution of the eye as follows. He maintains that the eye did indeed evolve in small increments, but this is a problem only if we ignore the fact that each eye evolves from earlier versions of itself that function effectively, but not perfectly. Hence, according to Dawkins, any mutation that improves the lens, the retina, the optic nerve, or brain would be selectively reinforced by natural selection. But, says Dawkins, such a change is likely to be small because large changes would affect vision adversely. To support his argument Dawkins appeals to the analogy of changes in a microscope that is slightly off-focus. If we shift the microscope viewing tube up or down by a very small distance there is a 50–50 chance of affecting positively or negatively the resolution of the microscope. However, a large movement either way would always have an adverse effect on the resolution. Similarly, so says Dawkins, many small mutations in the eye of an organism—some affecting the lens, some the retina, some the optic nerve—each of which is subjected to natural selection, can add up over long geological periods to make the eye more complex and perfect as an organ of perception. Hence, the recipe for producing a working eye, argues Dawkins, improves through various small but cumulative changes over long periods of time.

But Dawkins explanation of the evolution of the eye is really not consistent with his recipe view of the genetic code. He implicitly assumes that any random mutation affects one organ at a time—say the eye lens, but not the retina, optic nerve, or brain. If this were the case then we would indeed expect a gradual step by step selection of genes that affect various parts of the evolving perceptual system. But such an account presupposes the blueprint conception of genes that Dawkins rejects. His recipe view requires us to suppose that every mutation affects the organism as a whole since it involves a change in the recipe. However, this would create another problem for his viewpoint. If small mutations have a 50–50 chance of improving or degrading the functioning of any particular organ, then any small mutation in a recipe rule, which has a 50–50 chance of improving the eye, say, would also have a 50–50 chance of improving the optic nerve or the brain. This, in effect, would mean that any mutation would affect about 50% of the organs adversely and 50% advantageously. In such a situation natural selection can hardly bring about improvements—and if it did the process would be even slower than Dawkins supposes (in spite of his appeal to long periods of time). What Dawkins gains by adopting the recipe model—the explanation of holistic changes across the whole organism—he now loses because he has to face the problem that any random change in his recipe will affect about half of the organs in the organism positively and the other half adversely.

However, if we recognize that a small mutation in the genotype could result in coordinated changes of the lens, retina, optic nerves and brain—precisely the sort of simultaneous and integrated change that Dawkins precludes because of his recipe view of the genetic code, then the problem of accounting for the eye can be resolved. In order to appreciate this point let us consider in detail the stages through which the human eye develops in the process of embryogenesis, and how its links with the optic nerve and brain are established. In their description of the process of human embryogenesis the developmental biologists Ronan O'Rahilly and Fabiola Muller divide the prenatal life of the human embryo into two periods. First, we have the embryonic period proper, which occurs over the first eight weeks, involving histogenesis, when tissues form, and organogenesis, when the organ systems develop. Second, we have the fetal period which ends in birth when development shifts from the processes of tissue and organic differentiation to processes of growth. O'Rahilly and Muller divide the 8 weeks of embryogenesis into 23 developmental stages based on morphological criteria. The first stage of fertilization is followed by the second stage, lasting one-and-half to three days, in which the fertilized egg divides into 2 to 16 cells. Differentiating and multiplying through successive stages, and about the 57th day, the embryo reaches the 23rd stage, by which time its head, limb, eyes, ears, and other organs are fully formed (O'Rahilly and Muller 1992).

Let us now look more closely at the stages through which the human eye develops during this process of morphogenesis.[31] The first visible indication of the eye occurs after twenty-two days in the tenth stage of development. The eye starts as a differentiation of the neural fold of the forebrain called the optic *sulcus*. This deepens over few days to become the optic vesicle. By this time the right and left optic vesicles can be seen to be in communication with the cavity that will become the future third ventricle of the forebrain (*diencephalon*). Around the 32nd day the optic cup and lens pit form (stage 14), and by 33 days (stage 15) a small portion of the optic stalk, which will become the future optic nerve, has formed. The retina begins to develop over the next week, and continues to do so until about day 51 (stage 20). After 57 days (stage 23) the eye lens, the viscous body and the pigmented layer develop, and the cornea consolidates itself in the postovulatory weeks. (O'Rahilly and Muller 1992: 294)

The most significant feature of the morphogenetic development of the eye is that it grows out of the brain, indeed as a specialized offshoot of the brain, with the optic nerve, retina and eye lens emerging in that order. Moreover, each stage of the development of the eye proceeds from a context set by its earlier stage. It suggests that we do not need to account for the evolution of the eye by assuming many distinct coordinated mutations in the genotype which separately influence the structure of the optic centers of the brain, the optic nerves, the retina, the eye lens, and so on. Indeed, since the influence of any single mutation that affects the eye lens has to be mediated through the brain and optic nerves, it would automatically influence their development as well. Every mutation that affects the eye lens would also bring about changes in the brain, the optic nerve, and the retina. Moreover, it is precisely such correlated changes that are subjected to natural selection.

The above example exemplifies the crucial difference between Dawkins' view of the genetic code as expressing recipe rules and the view that it carries recursive rules for cell multiplication. Dawkins assumes that natural selection operates on separate small variations in the eye lens, retina, optic nerve, or brain that result from small mutations at the genetic level so as to bring about a coordinated change across these systems. By contrast, what actually occurs is that natural selection operates on large integrated and coordinated changes in the eye lens, retina, optic nerve, and brain in the phenotype resulting from small mutations in the genotype. Dawkins requires additional auxiliary assumptions, such as appeal to inordinately long time frames, to explain the integrated structure of the changes involved. By contrast, seeing the genotype as carrying recursive rules shows natural selection operates on coordinated macro-variations in the phenotype resulting from micro-mutations in the genotype. Hence, by allowing us to recognize that changes in the phenotypic variations can be large and integrated, even when the genotypic variations are small and isolated, the view of the genetic code as carrying recursive rules allows us to see how complex organs and organisms can evolve over much shorter periods than Dawkins thinks is required.

Let us now examine the third class of phenomena treated by Dawkins as especially problematic for Darwin's theory that evolution occurs by means of small random mutations subject to natural selection. These are traits in organisms—exemplified by so-called overshoot phenomena, or what Dawkins refers to as explosions or spirals—that seem to actually handicap them in their habitats. Such overshoot phenomena arise when an evolutionary trend continues far beyond the point where it bestows an advantage to the organism, and even to the point where it becomes a hindrance. The antlers of the Irish elk furnish one such example. These were so large that they weighed nearly a ton and spanned twelve feet. It is very difficult to consider them as bestowing an advantage to the animal, since they must have greatly reduced its mobility and freedom of movement.

Darwin had proposed sexual selection as a mechanism to explain such phenomena, and Dawkins follows him in this regard.[32] According to Dawkins large antlers not only provide an advantage in the combat for females during the mating season, but also result from the fact that females find them attractive. Hence Irish elk with overgrown antlers, which can be deemed secondary sexual characteristics, are likely to be more reproductively successful. But others have objected to the notion that we can explain the exaggerated horns by simply treating them as a trade-off between mobility and reproductive success. For example Taylor writes:

> The fact is when deer fight seriously they fight with their feet. In antler fights the antlers often become locked and both animals die. Antlers are thus a disadvantage to the species. (Taylor 1983: 27)

However, treating the genotype as a carrier of recursively applied rules that guide the morphogenetic process that forms an organism precludes any need to invoke sexual selection to explain overshoot phenomena. It now becomes possible to understand why the size of the antlers may become exaggerated simple because they are correlated with another trait such as the size of the feet. This would occur if the genes that affect antler size also affect feet-size because they are correlated in the process of morphogenesis, as we saw genes affecting the eye lens also affecting the cornea, retina, and optic nerve. In such a situation if increasing feet length is being selected, because it gives an advantage to the animal, then natural selection operating to increase feet size would also increase the size of the antlers as an incidental byproduct. Such a selection process would continue so long as the cost incurred by the increase in antler size is more than balanced by the advantage obtained as a result of larger foot size. Taken in isolation, the increase in antler size appears disadvantageous; but taken as a whole we can see large antlers as the price the animal pays to obtain larger feet useful in combat.

The role of correlated changes of this kind in evolutionary history has been noted by Varela et al. who write:

> The fact that the presence of a gene does not result in the manifestation of an isolated trait, except in a few remarkable cases (such as eye color) is known to biologists as linkage and pleiotropy … Pleiotropy provides obvious difficulties for adaptationism. How can a gene be selectively optimized if it has multiple effects, which need not increase fitness in the same manner or even in the same direction? Selection might push to decrease the frequency of a certain gene but pleiotropy, on the other hand, might push to increase or maintain the gene. The net result is some compromise that cannot be described as simply the result of selective pressures. (Varela et al. 1991: 188–189)

The same explanation could apply to many other overshoot phenomena where an evolutionary trend appears to persist far beyond the point of utility to an organism. The long dangling tail of the peacock which handicaps it in both flying and walking; the canine teeth of the smilodon[33] that are so large that they prevented some of them from closing their mouths, and in others had not only increased in size but also curved around so that they become useless as a defensive or offensive weapon. All these phenomena suggest an evolutionary momentum that seems to have carried the development of an organ way beyond any adaptive value.

Such overshoot phenomena are normally explained by appealing to sexual selection—the preference of females for males with long tails, or curved long tusks, and so on. Dawkins, in fact, appeals to sexual selection to explain overshoot phenomena. Using the tail of the African long-tailed widow bird he writes:

> Tails have an important job to perform in flight, and a tail that is too long or too short will decrease the efficiency of flight. Moreover, a long tail costs more energy to carry around, and more to make it in the first place. Males with 4-inch tails might well pull the female birds, but the price the males would

> pay is their less-efficient flight, greater energy costs and greater vulnerability to predators. We can express this by saying that there is a *utilitarian* optimum tail length, which is different from the sexually selected optimum: an ideal tail length from the point of view of ordinary useful criteria; a tail length that is ideal from all points of view apart from attracting females.
>
> Should we expect that the actual average tail length of males, 3 inches in our hypothetical example, will be the same as the utilitarian optimum? No, we should expect the utilitarian optimum to be less, say 2 inches. The reason is that the actual average tail length of 3 inches is the result of a compromise between utilitarian selection tending to make tails shorter, and sexual selection tending to make them longer. (Dawkins 1986: 204–205)

It is evident that Dawkins uses sexual selection as an auxiliary hypothesis to explain what he cannot explain solely by appeal to utilitarian selection. Yet such an auxiliary hypothesis becomes unnecessary once we recognize pleiotropy—namely, that the same gene could cause multiple effects and, therefore, by being selected for an effect that makes the organism more adaptive in one regard, it could also develop a correlated effect in the phenotype that is disadvantageous.[34] Using the above example, we can say that the longer tails that are maladaptive are the outcome of a gene that affects both tail length and another trait in the phenotype being selected for utilitarian reasons. The actual tail length is the result of a compromise between the increase in utility brought about by the correlated trait and the loss of utility due to increased tail length. There is no longer any need to invoke sexual selection as an auxiliary assumption to complement utilitarian selection—utilitarian adaptation alone suffices to explain both the evolution of adaptive traits in an organism, and the sometimes correlated incidental development of maladaptive traits, provided we recognize that genes guide organogenesis through recursively applied cell genetic rules that control the morphogenetic process.

The above examples suggest that the evolution of phenotypes can be explained from two mutually exclusive but equally necessary perspectives, both of which are needed to give a complete account of the phenomena observed. On the one hand, we can explain the evolution of phenotypes in terms of genotypic changes or variations. These are molecular changes involving small, isolated, incremental steps over time. On the other hand, we can explain their evolution from the point of view of adaption to the environment. Hence when biologists speak of organic variations they can do this from two quite different perspectives—a genotypic perspective in which variations arise randomly and cumulatively, and a phenotypic perspective that could change in disjunctive steps. Moreover, successful genotypic variations can be explained in terms of molecular changes in the genetic code, but their correlated phenotypic variations have to be explained functionally in terms of how they adapt organisms to the environment. It is only by adopting Bohr's biological complementarity view of molecular and functional explanations to deal with genotypic and phenotypic variations respectively that we can come to comprehensively understand the complete pattern of organic evolution.

The complementarity standpoint of combining notions of genotypic and environmental perspectives to explain phenotypic traits and behavior, analogous to the particle and wave perspectives in atomic physics, enables us to avoid postulating auxiliary hypotheses, over and above the notions of genetic random mutations and environmental natural selection. We do not need to appeal to gaps in the archeological record to explain discontinuous changes across species revealed in the record—they would be expected on this view since small genotypic variations can result in large phenotypic changes. We don't have to invoke inordinately long time periods to explain the emergence of "organs of extreme perfection"—natural selection can operate on coordinated macro-variations over a number of organ systems in the phenotype that arise from tiny variations or micromutations in the genotype. Neither do we need the auxiliary assumption of sexual selection to accommodate overshoot phenomena—natural selection suffices since selection of one property for a phenotype may overdevelop another property. Hence, the complementarity of the molecular and functional perspectives recommended by Bohr not only illuminates developmental biology, but also casts light and resolves problems in evolutionary biology.

2.4 Geology and Geophysiology: Complementary Perspectives on Ecosystems

Ecology is another area of biology in which grown properties play a significant role—a fact that often gets overlooked when a mechanical vision is extended to understand ecological systems. The important role of grown properties in ecosystems suggests that the epistemology of complementarity can also significantly illuminate this area of biology. However, the details of the application are likely to be different because ecological processes do not develop and evolve through processes of morphogenesis and organic evolution in the way that organisms do. In particular, ecosystem growth and evolution is not directed by a genetic code. Instead the major regulating factor in ecology is what we have earlier termed "the position effect". Indeed, the absence of the constraining influence of genes makes the influence of positional factors—environmental factors—in ecological processes even more significant.

The position effect is most evident when a disturbed ecosystem returns back through a number of developmental stages of ecological succession to either its original equilibrium or a new one. Though such processes have been studied since the beginning of the twentieth century, a generally acceptable theory of how ecosystems reach a stable climax has yet to be developed. There are many controversies concerning ecological succession. Debates continue about whether later species replace earlier species, or co-exist with them; whether earlier species modify the environment to exclude other species or make it conducive for them; the extent to which successional species arise from propagules (seeds, root fragments, whole plants) remaining in the soil and migrant propagules from elsewhere; whether the final climax ecosystem is undefined in advance or crucially determined by the initial distribution of propagules; and even whether an ecosystem ever stabilizes or goes through cyclical and continuous change (Luken 1990: 2–6).[35]

In spite of these divergent views it is generally conceded that succession is shaped by the adaptability of organisms, both flora and fauna, to their local context in the ecosystem. Each organism develops in response to its local environment, and then becomes a part of the environment in which others develop. It grows and survives in the context of others, and itself contributes to the environment to which others have to adjust. Hence, as in the case of morphogenesis where cells grow in dependence on the context of other cells and themselves become the context for the growth of new cells, the position effect also plays an important role in ecological succession. It explains why a disturbed ecosystem changes in a jerky fashion over time, remaining stable for some periods and then suddenly mutating to accommodate a new spectrum of species, as conditions develop to impede or facilitate different kinds of plants and animals. This understanding of ecosystems as conditioned by the position effect can even be extended to the largest ecosystem we know—the biosphere as a whole. This influence of the position effect is what lies at the heart of the so-called "Gaia hypothesis" in approaching ecological processes and explains why the complementarity principle has applicability here.[36]

The Gaia hypothesis was first developed by scientist and environmentalist James Lovelock. It treats the biosphere as a superorganism that is a self-regulating system. Lovelock's hypothesis was supported and co-developed by biologist Lynn Margulis who is, however, somewhat more skeptical about the organic metaphor. Margulis formulates the hypothesis as follows:

> The Gaia hypothesis states that the earth's surface conditions are regulated by the activities of life. Specifically, the earth's atmosphere is maintained far from chemical equilibrium with respect to its composition of reactive gases, oxidation-reduction state, alkalinity-acidity, albedo, and temperature. This environmental maintenance is effected by the growth and metabolic activities of the sum of the organisms, i.e., the biota.[37]

The Gaia hypothesis serves to explain what might otherwise appear to be a miraculous series of coincidences that somehow happened to make life on earth possible. The earth has just the right proportion of oxygen in relation to the other gases needed to support life—if the proportion were more the forests would become easily combustible; if less, life would not be possible. The amount of carbon dioxide also appears to be designed to maintain the earth's temperature at a level suitable for life—more carbon dioxide would make the temperature too high due to the greenhouse effect; less, and the photosynthetic activities, which produce food for plants (and ultimately all life), would not be feasible. The salt concentration of the oceans is maintained at a level acceptable to living things through the removal of salts, washed down by rainwater and rivers into the oceans, by complex processes. Moreover, since solar luminosity has increased over the eons life has existed on the earth, we might have expected a rise of temperatures on the earth beyond levels amenable to supporting life. This has not happened because of other compensating processes that gradually reduced the proportion of carbon dioxide in the atmosphere, so as to reduce the greenhouse effect and maintain temperatures within the range suitable for life.

According to the Gaia hypothesis all of these regulating processes are managed by the earth's living organisms. Hence, argues Lovelock, the biosphere as a whole—or Gaia—behaves like a living organism whose internal processes are continually regulated by living things that are a part of it. Many of the geological properties of the earth, which had originally been taken to provide conditions accidentally suited to life are seen by him to be the outcome of the activities of living organisms—activities that create the very conditions that sustain life. He concludes that Gaia is "a self-regulating entity with the capacity to keep the planet healthy by controlling the chemical and physical environment."[38]

This leads Lovelock to recommend that we can advance understanding by adopting a geophysiological orientation in the earth sciences—one that would encourage us to explain many features of the earth's different ecosystems in terms of their interdependence and connectivity with each other in the way the organs of a living organism are linked to each other. He writes:

> Geophysiology reminds us that all ecosystems are interconnected. By analogy, in an animal, the liver has some capacity for the regulation of its internal environment, and its liver cells can be grown in the isolation of tissue culture. But neither the animal nor its liver can live alone; they depend upon their interconnection.[39]

On the Gaian geophysiological model individual living organisms can be treated as parts of a larger living organism—Gaia.[40] Lovelock visualizes the situation as follows:

> As we move in towards the earth from space, first we see the atmospheric boundary that encloses Gaia; then the borders of an eco-system such as the forests; then the skin or bark of living animals and plants; further in are the cell membranes; and finally the nucleus of the cell and its DNA. If life is defined as a self-organizing system characterized by an actively sustained low entropy, then viewed from the outside of each of these boundaries, what lies within is alive. (Lovelock 1989: 27)

Lovelock's proposal is intriguing. Although he is not rejecting the geological approach of understanding biospheric processes in physical and chemical terms, he suggests that it can be enriched by a complementary geophysiological approach that treats the earth as an organic system. Adopting Lovelock's view would extend Bohr's suggestion that our understanding of biological processes should combine molecular and functional approaches. The complementarity perspective would now be extended to include ecological, along with developmental and evolutionary processes.

However, the Gaia hypothesis faces one obstacle. The notion that the earth is a superorganism has met with wide resistance because it appears teleological. How can organisms and micro-organisms act co-operatively, in the way of organ-systems in a living body, to modify or shape features of the global environment? This is analogous to the problem we confronted in morphogenesis—how can individual cells be sensitive to the global context of the organism so that they "know" whether they are to become a skin cell, eye cell, or blood cell?

In the case of morphogenesis, we avoided teleology by explaining the process as guided by the position effect. No plan was contained in the DNA code for the organism; the organism was merely the outcome of instructions in the genetic code that told each cell how to behave in its local context—what sort of properties to acquire, and how to produce new cells. This enabled us to see the growth and multiplication of cells without attributing any teleology that guided them. Similarly, if we can show that the global equilibrium that sustains life in the biosphere is itself the unintended and incidental outcome of organisms responding to their local contexts, then it would be possible to give an explanation of Gaian self-regulation without teleological presumptions. In fact it is this kind of sensitivity to local environment that Lovelock invokes. It enables him to get around the objection that his theory is teleological because it assumes organisms deliberately set out to cooperatively create a global environment suitable for their local needs.

In order to show how it is possible for a large number of local adaptations to lead to a global adaptation, without making teleological assumptions, Lovelock offers us his model of Daisyworld (Lovelock 1989: 35–41).[41] He considers an imaginary planet about the size of the earth orbiting, in a similar fashion, a star of the same mass and size as the sun. He assumes that daisies of different shades of color—dark, light, and neutral—inhabit the planet. He further presupposes that daises will not grow at temperatures below 5 degrees Celsius or above 40 degrees Celsius, and that they grow best at 20 degrees Celsius. The temperature of the planet as a whole is determined both by the luminosity of its sun and the planet's albedo—that is, the average color of the planet. If the albedo is low, that is, the planet is dark, it would absorb more heat from sunlight and be warmed; if the albedo is high then it will be lighter in color, reflect more heat, and become cooler.

Lovelock then proposes that we assume that the star was less luminous in the distant past. Then temperatures would have been warm enough at

5 Degrees Celsius for daisies to grow only near the equator of Daisyworld. At this time dark daisies would be favored since they would absorb more sunlight and be more easily warmed above 5 Degrees Celsius. By contrast, white daisies, because they reflect light, would cool below 5 Degrees Celsius and die. Soon dark daisies would spread across the planet as they slowly warmed more areas of it above 5 Degrees Celsius until they end up colonizing the planet. But as they spread and the planet warms, white daisies would also be able to flourish. Indeed, when the planetary temperature reaches 20 Degrees Celsius any further increase in black daisies would raise the temperature above this optimum level for daisies. Thus, beyond this temperature less black daisies would survive and more white daisies would begin to grow, since they would warm up less by reflecting starlight. The increase in white daisies would also cool the planet by reflecting off more radiation. Vice versa, a fall below the optimal temperature of 20 Degrees Celsius would result in more black and fewer white daisies.

We know today that our own sun has increased in luminosity by about 30 percent since life began on earth. Suppose we assume that the same happens with the sun of Daisyworld. Lovelock argues that this would cause the proportion of white daisies to increase until finally even the whitest crop of daisies cannot keep the planet below the critical temperature of 40 Degrees Celsius. At this point Daisyworld would die. Lovelock uses this example to explain, without invoking teleological notions of foresight and planning, how a *global* property of the environment, in this case the temperature of Daisyworld, can be regulated by its biota over a wide range of solar luminosity simply by responding to *local* conditions.[42]

However, a critic could still attack Lovelock's Daisyworld argument as follows. There is no guarantee that the local modifications an organism makes to the environment, in order to enhance conditions for itself, will always contribute to altering the global environment to benefit all organisms in it. After all, there are polluting organisms that enhance conditions for themselves by creating destructive consequences for others. There is no guarantee that each organism responding to its local context would, as a byproduct, contribute to producing global conditions necessary for the welfare of all.

However, Lovelock's theory becomes more credible if we consider the following argument. There are a narrow range of conditions that are needed for all organisms to survive on earth. For such conditions any improvement an organism makes for itself will also, as an incidental byproduct, improve the conditions for all.[43] To see this clearly consider Lovelock's description of the environmental conditions needed by all life:

> [We must recognize] the existence of constraints or bounds that establish the limits of life. It can be too hot or too cold; there is a comfortable warmth in between, the preferred state. It can be too acid or too alkaline; neutrality is preferred. Almost all chemicals have a range of concentrations tolerated or needed by life. For many elements such as iodine, selenium and iron, too much is poison, too little causes starvation. (Lovelock 1989: 40)

We have seen that in Lovelock's Daisyworld the local rising temperatures caused black daisies to die locally and to be replaced by white daisies. Although all the responses made by the daisies are to local conditions, these local responses also happen to contribute to maintaining global conditions on the planet for daisies. We may extend this argument to many of the boundary conditions that make life possible on earth since these are more or less the same for all living things. Historically these conditions had been seen as accidentally present on earth before life became possible. By contrast, Lovelock argues that the conditions themselves are created and sustained by living things—largely by being incidental outcomes of the local behavior of the sum of biota on earth.

Clearly, any local adaptation that moderates temperature by reflecting heat out of the biosphere, or neutralizes acidity or alkalinity, or converts harmful chemicals into less lethal ones, would also help to make the global environment more amenable to life in general. But since each organism is only responding to its local environment, we cannot deem the outcome to be achieved on purpose—it is not a teleological outcome.

Nevertheless, an extended chain of such local adaptations can result in highly integrated responses of macro-systems in the biosphere in the same way that the responses of individual cells in an organism, each to its local context, can produce highly integrated behaviors of organ systems. For this reason Lovelock suggests that it is heuristically fruitful to adopt a geophysiological perspective on the biosphere. Such a Gaian perspective would avoid assuming that a set of miraculous coincidences brought about just those narrow geological conditions—temperatures not so high that water would boil or so low it would freeze; the right atmospheric composition of gases, appropriate levels of marine salinity and so on—which make life possible on Earth.

Lovelock's recommendation that ecological processes can be illuminated by a geophysiological perspective leads him to argue that the earth is similar to an organism. He proposes that "Gaia as the largest manifestation of life differs from other living organisms of the earth in the way that you or I differ from our population of living cells". (Lovelock 1989: 41) However, there are many differences between Gaia and an organism such as a plant or an animal. First, Gaia does not reproduce itself by multiplication even if it continually reproduces its parts to sustain itself. Second, Gaian processes are not guided by a genetic code—only some of its component parts carry genetic codes. Third, although Gaia evolves, it does not do this as a result of natural selection. Being an entity that is not mechanical because of the importance of grown properties that condition its structure and behavior; and one that is not organic because it neither possesses a genetic code nor is subject to natural selection, it is best described as quasi-organic.[44]

As a quasi-organic system Gaia self-regulates itself through the position effect. Such a quasi-organic view justifies Lovelock's recommendation to combine the geophysiological approach with the geological approach stressing the physics and chemistry of the biosphere. Such a combined approach would be better placed to recognize ecological linkages across diverse ecosystems in the biosphere. Following Lovelock we would treat local ecosystems and individual organisms within Gaia as functionally analogous to the organ systems and cells of an animal or plant. We would then approach understanding them in terms of the role they play in contributing to the welfare of the biosphere as a whole.

The self-regulation of Gaia through the position effect provides the key for extending Bohr's notion of the molecular and functional points of view as mutually exclusive but complementary perspectives that can be adopted to understand ecological processes. It makes it possible to treat the geophysiological approach that sees ecosystems functionally in terms of their role in the biosphere, as complementary to the geological approach of understanding ecosystems by looking at the mechanical interactions of their component parts. This would, in effect, extend Bohr's biological complementarity to help articulate our understanding of the structure and evolution of ecosystems.

Extending biological complementarity to the biosphere would also have important implications for conservation practices in general. It suggests that the biosphere as the habitat of organisms is itself sustained by the organisms within it. Moreover, these organisms themselves develop and survive in the larger context of their environment. It is the organism–habitat interactions that generate organisms in their habitats, and preserve habitats by means of processes carried out by the organisms within them. This means that we cannot conserve species by setting up gene-banks or storing specimens in zoos. The position effect reminds us that habitats, and not merely genes, are crucial carriers of developmental information, so that the strategy of only preserving genes, without also protecting the habitats in which they can actualize themselves, could be a delusion leading us into complacency and a false sense of achievement.

Moreover the Gaia hypothesis sensitizes us to the role of organisms in generating and preserving their habitats. It is the gene–habitat interaction that reproduces an organism with its full repertoire of capacities, as well as contributes to the habitats necessary for its survival. Indeed, without appropriate programs today to shield habitats and the biosphere—the largest habitat of all—from degradation we may find ourselves with elaborate storehouses of useless genetic information because large chunks of other equally necessary information have been eliminated through habitat degradation or destruction.

We have seen that the pervasive presence of grown properties in nature is what makes it possible to extend the notion of biological complementarity—the complementarity of the molecular and functional points of view—to morphological, evolutionary, and ecological processes. The ubiquity of grown properties also shapes the way we come to experience nature. This dependence of psychological responses to grown properties in nature explains, as we find, the parallels Bohr noticed between the philosophical implications of quantum theory and Buddhist epistemology. Thus, psychological complementarity and parallels to complementarity in Buddhist thought are not unrelated. Indeed, in the next chapter, we show that Bohr's extensions of complementarity into psychology can not only pave the way for explaining epistemological parallels in Indian philosophy to complementarity, but also show how insights from Buddhist epistemology can enrich the epistemology of science.

We find biological complementarity also has implications for how we should manage our economic relations with nature in a manner that not only meets our needs but also protects the self-regulating processes that maintain the integrity of ecosystems. In particular our interactions with nature must take into account the preservation of what environmental economists have referred to as 'natural capital'. In this regard Bohr's extension of the principle of complementarity into biology and his discovery of similar notions in Daoist philosophy, are significant. Indeed, in the penultimate chapter of this study we see how Daoist insights can be made to inform economic theory in order to articulate more sustainable relations with nature.

Notes

1. For a historical discussion of these controversies see Jammer (1974), Sachs (1988) and Peat (1990). There is also the lively debate between Einstein and Bohr best documented by Schilpp (1949). For a recent, but tendentious, view on these debates see Beller (1999).
2. Heisenberg also rejected Einstein's attempt to dictate what nature should be like. In particular Heisenberg did not agree with Einstein that his uncertainty principle implied the incompleteness of the quantum theory. Einstein's dissatisfaction is expressed in a letter to Max Born in 1926 when he writes: "Quantum mechanics is certainly imposing. But an inner voice tells me that it is not yet the real thing. The theory says a lot, but does not really bring us any closer to the secret of the "old one." I, at any rate, am convinced that He does not throw dice." Letter to Max Born (4 December 1926); *The Born-Einstein Letters* (translated by Irene Born) (Walker and Company, New York, 1971).
3. Einstein et al. (1935).
4. See Bell (1964). This article is reprinted in Bell (2004); see Chap. 2.
5. For a comprehensive discussion of the early experiments see the review articles by Clauser and Shimony (1978) and Redhead (1987). A more recent account of such experiments, and their implications for understanding nature, can be found in Steward (2011).
6. Aspect et al. (1981, 1982).
7. For arguments that the logic of quantum physics itself may prevent us from ever setting up a loophole-free test, see Santos (2005) and Gill (2003).
8. This is taken with minor changes from Mansfield (1989: 380) who follows Mermin's (1981: 940) formulation of the experimental situation.
9. For a derivation of this proportion using only simple mathematics see Mansfield (1989), pp. 388–391.
10. James T. Cushing, "A Background Essay" in Cushing and McMullin (1989), p. 9.
11. Niels Bohr, "Discussions with Einstein on Epistemological Problems in Atomic Physics," in P.A. Schilpp (1949) pp. 209–210.
12. For a discussion of the various historical factors that led to Bohr's views being identified with positivism, not least that Bohr opposed Einstein who opposed the positivists, see also Folse (1985), pp. 18–27.

13. Mach and Avenarius were nineteenth century positivists essentially concerned with rejecting the atomic theory of matter. It is strange that Bohr should be linked to positivist views because his epochal work that opened the door to later developments of the quantum theory was based upon a planetary model of the atom.
14. Folse (1985, p. 262) stresses the same point. One may also invoke Ian Hacking's distinction to argue that the quantum object exists as a real entity prior to measurement, but that the property measured arises within the context of measurement. See Hacking (1983) For a critical appraisal of entity realism see Resnik (1994), Shapere (1993), and Gelfert (2003).
15. Held would contest this conclusion. He writes that in his mature years Bohr relinquished the notion of wave–particle complementarity and reinterpreted the complementarity of space-time and causal descriptions in response to the Einstein's critique of quantum theory. See Held (1994), pp. 871–2. However, he provides no evidence for this and Bohr's emphasis on space-time and causal descriptions may simply reflect his attempts to deal with the EPR paradox where wave–particle complementarity was not the immediate issue.
16. For a history of genetic theory and the changing paradigms associated with it see John C. Avise (2014).
17. For recent references of phenotypic plasticity, see West-Eberhard (2003), Price, Qvarnström and Irwin (2003), Barker (2008), Barker, Desjardins and Pearce (2014), and Fordyce (2006).
18. The same point is developed by Meyerowitz (1994) who examines the genetics of flower development and argues that flower cells differentiate into flower organs by responding to positional information. See also Wolpert and Tickle (2011).
19. For more details of the one-many relations between genotype and phenotype, see Griffiths et al. (2000).
20. Although Dawkins first published his book in 1986, the general approach it develops has not been abrogated by Dawkins as a result of critiques of his reductionist views by punctuationists, who assign a greater role to the organism's interactions with the environment in the evolutionary process. However, in a later 1996 edition of the book Dawkins notes that he reached a greater appreciation of embryological constraints on natural selection as a result of his work with computer simulation models. His more recent work *The Greatest Show on Earth*: *The Evidence for Evolution* (2010) has shifted focus to pay attention, not to scientific critics of his views, but those who oppose evolution on creationist grounds.

21. It is noteworthy that Dawkins view is designed to challenge the theory proposed in 1972 by two paleontologists Niles Eldredge and Stephen J. Gould who published a paper proposing that the geological record shows that most species remain in a state of stasis for long periods with little evolutionary change. When change occurs it is a geologically rapid process that brings about branching speciation. Their theory of punctuated equilibrium suggests that species split into two different species in a sudden process rather than through gradual transformation.

    By contrast, Dawkins argues that evolution does not proceed at a uniform rate—what he calls "constant speedism"—but at variable speed depending on selective pressures. He contrasts such continuously variable speedism with Eldredge and Gould's thesis that evolution exhibits what he terms "discrete variable speedism"—namely that it alternates between stability and bursts of rapid change. Consequently, Dawkins explains the apparent gaps in the fossil record as the outcome of migratory events. For him evolutionary changes occurred gradually elsewhere, but give the illusion of rapid change unless we unearth fossil evidence for this process.

    The view we are proposing here is more in tune with Eldredge and Gould's organism-centered view than the gene-centered perspective of Dawkins. Particularly significant has been Gould's notion that the unit of selection is the phenotype and not the genotype—the phenotype is what interacts with the environment at the interphase of natural selection. Hence, Gould maintains that genes do not have direct visibility to natural selection. See "Caring Groups and Selfish Genes", Gould (1990), pp. 72–78. Kim Sterelny (2007: 83) describes Gould's position as proposing "gene differences do not cause evolutionary changes in populations, they register those changes." See also Gould (2002).
22. Darwin writes:

    > Nevertheless the simplicity of the view that each species was first produced within a single region captivates the mind. He who rejects it, rejects the *vera causa* of ordinary generation with subsequent migration, and calls in the agency of a miracle. (Darwin 1859: 352)

23. Indeed Darwin did not separate these two types of variations. He was solely concerned with phenotypic variations, since he was not aware of Mendel's discoveries. However, Darwin did say that "I am strongly inclined to suspect that the most frequent cause of variability may be attributed to the male and female reproductive elements having been affected prior to the act of conception." However, he saw small changes in the "reproductive elements" only as having small effects in the organism which develops (Darwin 1859: 8).
24. Niche-selection can be seen as an example of the Baldwin effect and genetic assimilation. See Crispo (2007). However, recent work emphasizes the

role of niche construction followed by natural selection. See Barker (2008) and Odling-Smee, Laland, and Feldman (2003).

25. Hox genes, which determine the basic structure and orientation of phenotypes, illustrate the possibility of such homeotic mutations. These genes determine the placement of segment structures, including legs, antennae, and wings, during the early stages of embryonic development in animals. They have been particularly well-studied in the fruitfly, *Drosophila melanogaster*. See Hunt (1998).
26. See Lewis, E.B. (1978). For an update survey of such research see Heffer, A. & Pick, L. (2013).
27. Quoted in Koestler (1967) pp. 141–142.
28. Such laws of form have been proposed by Arthur Koestler (1967: 142) based on the work of D'Arcy Thompson's study *On Growth and Form*. However, Thompson had argued that structuralism, not laws of form, governed the morphology of species. In particular, he used many examples to illustrate important correlations between mechanical processes and biological forms. For example, engineering truss designs bore similarities to supporting structures in the hollow bones of birds, and drops of liquid falling into a viscous fluid gave rise to structures similar to jellyfish. He also explored parallels in the spiral structures of plants to the Fibonacci series. Most striking were his illustrations of how the forms of different species of animals could be seen as topological transformations of each other, such as the *Argyropelecus olfersi* and *Sternoptyx diaphana*. See Thompson (1942).
29. He wrote:

    > Can we believe that natural selection could produce, on the one hand, organs of trifling importance, such as the tail of a giraffe, which serves as a fly-flapper, and, on the other hand, organs of such wonderful structure, as the eye, of which we hardly as yet fully understand the inimitable perfection? (Darwin 1859: 171–172)

    The eye has evolved separately in many animal organisms from the time of the Cambrian explosion nearly 550 million years ago. See Land and Nilsson (2012). See also the journal *Evolution: Education and Outreach* Volume 1, Number 4 (2008). Its *Special Issue: The Evolution of Eyes* carries 26 articles by professional scientists about various aspects of evolutionary science of the eye.
30. Darwin's letter to Asa Gray, a Christian minister, about 1860. From "Quotations from Darwin" in *The Life and Letters of Charles Darwin*, Volumes I & II, edited by his son, Francis Darwin.Other examples of organs of extreme perfection that Darwin considered particularly problematic for the notion of evolution as proceeding through small cumulative changes are the ear and the circulatory system. See Darwin (1859), espe-

cially the section on "Organs of Extreme Perfection" in Chap. VI addressing difficulties for the theory.

31. O'Rahilly and Muller (1992), pp. 293–303. They give us a descriptive account of the way the human eye develops through a series of inductions shaped by the position effect. Over the last twenty years we have acquired a lot more understanding of the complex interplay between inductive signals provided by tissue–tissue interactions and cell-intrinsic factors that shape the morphogenetic emergence of the eye, yielding greater knowledge of both extrinsic and intrinsic determinants of the process. For more details on these advances see Cagan and Reh (2010), especially Chap. 3 by Sabine Fuhrmann on "Eye Morphogenesis and Patterning of the Optic Vesicle." See also Gasser, Cork, Stillwell and McWilliams (2014).
32. Some critics of the hypothesis of sexual selection have argued that many of the traits considered sexually attractive are really warning signals used by organisms to enhance their survival rather than attract mates. Indeed warning displays use the same repertoire of behavioral, visual, auditory and olfactory means invoked by supporters of sexual selection. These critics think that the advocates of sexual selection wrongly interpret warning displays as designed to meet sexual goals. See Ruxton et al. (2004).
33. The smilodon, popularly known as the saber-toothed tiger, is an extinct species that flourished in North and South America in the Pleistocene Age from about 2.5 million to 10 000 years ago.
34. For a review of the history of pleiotropy, see Stearns (2010).
35. For recent studies linking successional theory with ecological restoration projects, see Walker et al. (2007).
36. For a recent general and broad review of the social, spiritual and scientific dimensions of the Gaia hypothesis, see Ruse (2013).
37. Margulis and Hinkle (1988) p.11, also quoted in Joseph (1990) p. 86. See also Lovelock (1989) p. 19 for the formulation of the hypothesis.
38. Such self-regulation is essentially driven by the free energy available from sunlight. (Lovelock 1989: 31).
39. James Lovelock. "Geophysiology: a new look at earth science," in personal website http://www.jameslovelock.org/page36.html (Accessed 2016).

    Originally published in *Bulletin Of The American Meteorological Society* (1986 Vol. 67, No. 4) and in Dickinson (ed. 1987).
40. Although the term "geophysiology" was popularized by James Lovelock, the notion was anticipated nearly two centuries earlier by geologist James Hutton who had proposed that the earth should be considered a superorganism, which could be studied fruitfully by adopting the approach of physiology. Many succeeding scientists, such as the English biologist Thomas Henry Huxley, the Russian founder of bio-geochemistry, Vladimir Vernadsky, and the American plant ecologist and pioneer of vegetation

succession studies, Frederick Clements also saw plants and animals as part of a larger organism.

41. Although Lovelock's original model only involved daisies, there were later extensions of the Daisyworld model which included foxes, rabbits, and other species. It led to the unexpected discovery that there were improving effects on the entire planet when the number of species increased. These included better temperature regulation and a more robust and stable system even under perturbations. It showed that biodiversity has positive impacts on planetary conditions. See Lenton and Lovelock (2001), and Wood et al. (2008).
42. Thus Lovelock writes:

> In Daisyworld, one property of the global environment, temperature, was shown to be regulated effectively, over a wide range of solar luminosity, by an imaginary planetary biota without invoking foresight or planning. This is a definite rebuttal of the accusation that the Gaia hypothesis is teleological, and so far it remains unchallenged. (Lovelock 1989: 39)

43. This argument applies only to organisms living within the thin skin of the earth's biosphere. Of course there are other kinds of organisms such as those living in hydrothermal vents of tectonic plates that should be considered to live in a completely different ecosystem. See van Dover (2000).
44. For a recent exploration of the scientific, philosophical, and theoretical foundations of Gaia see Schneider et.al. (2008).

Bibliography

Aspect, Alain, Philippe Grangier, and Gerard Roger. 1981. Experimental Tests of Realistic Local Theories via Bell's Theorem. *Physical Review Letters* 47(7): 460–463.

———. 1982. Experimental Realization of Einstein-Podolsky-Rosen-Bohm Gedanken Experiment: A New Violation of Bell's Inequalities. *Physical Review Letters* 49(2): 91–94.

Avise, John C. 2014. *Conceptual Breakthroughs in Evolutionary Genetics: A Brief History of Shifting Paradigms*. Amsterdam: Academic Press.

Barker, Gillian. 2008. Biological Levers and Extended Adaptationism. *Biology and Philosophy* 23(1): 1–25.

Barker, Gillian, Eric Desjardins, and Trevor Pearce, eds. 2014. *Entangled Life: Organism and Environment in the Biological and Social Sciences*. New York: Springer.

Bell, John S. 1964. On the Einstein Podolsky Rosen Paradox. *Physics* 1(3): 195–200.

Bell, John S 2004. *Speakable and Unspeakable in Quantum Mechanics: Collected Papers on Quantum Philosophy*. Cambridge, MA: Cambridge University Press.

Beller, Mara. 1999. *Quantum Dialogue: The Making of a Revolution*. Chicago, IL: University of Chicago Press.

Bohr, Niels. 1935. Can Quantum Mechanical Description of Physical Reality Be Considered Complete? *Physical Review* 48: 696–702.

Cagan, Ross L., and Thomas A. Reh, eds. 2010. *Invertebrate and Vertebrate Eye Development*. San Diego, CA: Elsevier/Academic.

Clauser, J.F., and A. Shimony. 1978. Bell's Theorem: Experimental Tests and Implications. *Reports on Progress in Physics* 41: 1881–1927.

Crispo, Erika. 2007. The Baldwin Effect and Genetic Assimilation: Revisiting Two Mechanisms of Evolutionary Change Mediated by Phenotypic Plasticity. *Evolution* 61–11: 2469 2479.

Cushing, James T., and Ernan McMullin. 1989. *Philosophical Consequences of Quantum Theory: Reflections on Bell's Theorem*. Notre Dame, IN: University of Notre Dame Press.

Darwin, Charles. 1859. *The Origin of Species*. London: John Murray.

———. 1868. *The Variation of Animals and Plants Under Domestication*. New York: O. Judd.

Dawkins, Richard. 1986. *The Blind Watchmaker: Why the Evidence of Evolution Reveals a Universe Without Design*. Harlow: Longman Scientific & Technical.

Dickinson, Robert E., ed. 1987. *The Geophysiology of Amazonia: Vegetation and Climate Interactions*. New York: Wiley for the United Nations University.

Einstein, A., B. Podolsky, and N. Rosen. 1935. Can Quantum Mechanical Description of Physical Reality Be Considered Complete? *Physical Review* 47: 777–780.

Feyerabend, Paul. 1978. *Against Method: Outline of an Anarchistic Theory of Knowledge*. London: Verso.

Folse, Henry J. 1985. *The Philosophy of Niels Bohr: The Framework of Complementarity*. Amsterdam: Elsevier Science Publishers.

——— 2001. *Bohr's Conception of the Quantum Mechanical State of a System and Its Role in the Framework of Complementarity*. arXiv:quant-ph/0210075v1.

Fordyce, James A. 2006. The Evolutionary Consequences of Ecological Interactions Mediated Through Phenotypic Plasticity. *Journal Experimental Biology* 209: 2377–2383.

Gasser, Raymond, John Cork, Brian Stillwell, and David McWilliams. 2014. Rebirth of Human Embryology. *Developmental Dynamics* 243: 621–628.

Gelfert, A. 2003. Manipulative Success and the Unreal. *International Studies in the Philosophy of Science* 17: 245–263.

Gill, R.D. 2003. Time, Finite Statistics, and Bell's Fifth Position. *Foundations of Probability and Physics—2*, 179–206. Vaxjo University Press. arXiv:quant-ph/0301059v2.

Gould, Stephen Jay. 1990. *The Panda's Thumb: More Reflections in Natural History*. London: Penguin Books.

Gould, Stephen. 2002. *The Structure of Evolutionary Theory*. Cambridge, MA: Belknap Press of Harvard University Press.

Grant, Verne. 1963. *The Origin of Adaptations*. New York: Columbia University Press.

Griffiths, A.J.F., J.H. Miller and D.T. Suzuki, et al. 2000. Norm of Reaction and Phenotypic Distribution. In *An Introduction to Genetic Analysis*. New York: W.H. Freeman.

Gunaratne, R.D. 1986. Understanding Nagarjuna's Catuskoti. *Philosophy East and West* 36(3): 213–234.

Hacking, Ian. 1983. *Representing and Intervening*. Cambridge, MA: Cambridge University Press.

Hanson, Norwood Russell. 1958. *Patterns of Discovery: An Inquiry into the Conceptual Foundations of Science*. Cambridge, MA: Cambridge University Press.

Heffer, A., and L. Pick. 2013. Conservation and Variation in Hox Genes: How Insect Models Pioneered the Evo-Devo Field. *Annual Reviews in Entomology* 58: 161–179.

Held, Carsten. 1994. The Meaning of Complementarity. *Studies in History and Philosophy of Science Part A* 25(6): 871–893.

Hitching, Francis. 1982. *The Neck of the Giraffe or Where Darwin Went Wrong*. London: Pan.

Hunt, Paul (1998). "The Function of Hox Genes". In Bittar, E. Edward (ed.) (1968) *Developmental Biology*. Amsterdam Elsevier.

Jammer, Max. 1974. *The Philosophy of Quantum Mechanics: The Interpretation of Quantum Mechanics in Historical Perspective*. New York: John Wiley & Sons.

Joseph, Lawrence E. 1990. *Gaia: The Growth of an Idea*. London: Arcana, Penguin Books.

Koestler, Arthur. 1967. *The Ghost in the Machine*. Chicago: Henry Regnery Company.

Land, Michael, and Dan-Eric Nilsson. 2012. *Animal Eyes*. Oxford: Oxford University Press.

Lenton, Timothy M., and James E. Lovelock. 2001. Daisyworld Revisited: Quantifying Biological Effects on Planetary Self-Regulation. *Tellus* 53B: 288–305.

Lewis, E.B. 1978. A Gene Complex Controlling Segmentation in Drosophila. *Nature* 276: 565–570.

Lovelock, James. 1989. *The Ages of Gaia: A Biography of our Living Earth*. Oxford: Oxford University Press.

Luken, James O. 1990. *Directing Ecological Succession*. London: Chapmen and Hall.

Mansfield, Victor. 1989. Madhyamika Buddhism and Quantum Mechanics: Beginning a Dialogue. *International Philosophical Quarterly* 29: 371–391.

Margulis, Lynn and Gregory Hinkle. 1988. Biota and Gaia. *Abstracts of First Chapman Conference on Gaia Hypothesis.*

Mermin, David. 1981. Bringing Home the Atomic World: Quantum Mysteries for Anybody. *American Journal of Physics* 49: 940–943.

Meyerowitz, Elliot M. 1994, November. The Genetics of Flower Development. *Scientific American*: 40–47.

Odling-Smee, John, Kevin Laland, and Marcus Feldman. 2003. *Niche Construction: The Neglected Process in Evolution*. Princeton, NJ: Princeton University Press.

O'Rahilly, Ronan R., and Fabiola Muller. 1992. *Human Embryology & Teratology*, New York: Wiley-Liss.

Peat, David. 1990. *Einstein's Moon: Bell's Theorem and the Curious Quest for Quantum Reality*. Chicago, IL: Contemporary Books.

Price, T.D., A. Qvarnström, and D.E. Irwin. 2003. The Role of Phenotypic Plasticity in Driving Genetic Evolution. *Proceeding of the Biological Sciences* 270(1523): 1433–1440.

Redhead, Michael. 1987. *Incompleteness, Nonlocality and Realism: A Prolegomenon to the Philosophy of Quantum Mechanics*. Oxford: Oxford University Press.

Resnik, D. 1994. Hacking's Experimental Realism. *Canadian Journal of Philosophy* 24: 395–412.

Ruse, Michael. 2013. *The Gaia Hypothesis: Science on a Pagan Planet*. Chicago, IL: University of Chicago Press.

Ruxton, Graeme, Thomas Sherratt, and Michael Speed. 2004. *Avoiding Attack: The Evolutionary Ecology of Crypsis, Warning Signals, and Mimicry*. Oxford: Oxford University Press.

Sachs, Mendel. 1988. *Einstein versus Bohr: The Continuing Controversies in Physics*. La Salle, IL: Open Court.

Santos, E. 2005. Bell's Theorem and the Experiments: Increasing Empirical Support to Local Realism. *Studies in History and Philosophy of Modern Physics* 36(3): 544–565.

Saunders, Simon. 2005. Complementarity and Scientific Rationality. *Foundations of Physics* 35(3): 431–432.

Schilpp, P.A., ed. 1949. *Albert Einstein: Philosopher—Scientist*. Evanston, IL: Library of Living Philosophers.

Schneider, Stephen, James Miller, Eileen Crist, and Penelope Boston, eds. 2008. *Scientists Debate Gaia: The Next Century*. Cambridge, MA: The MIT Press.

Shapere, D. 1993. Astronomy and Antirealism. *Philosophy of Science* 60: 134–150.

Stearns, F.W. 2010. One Hundred Years of Pleiotropy: A Retrospective. *Genetics* 186(3): 767–773.

Sterelny, Kim. 2007. *Dawkins vs Gould: Survival of the Fittest*. Cambridge, UK: Icon Books.

Steward, Edward George. 2011. *Quantum Mechanics: Its Early Development and the Road to Entanglement and Beyond*. London: Imperial College Press.

Taylor, Gordon Rattray. 1983. *The Great Evolution Mystery*. New York: Harper & Row.

Thompson, D.W. 1942. *On Growth and Form*. Cambridge, MA: Cambridge University Press.

van Dover, Cindy Lee. 2000. *The Ecology of Deep-Sea Hydrothermal Vents*. Princeton, NJ: Princeton University Press.

Varela, Francisco, Evan Thompson, and Eleanor Rosch. 1991. *The Embodied Mind:Cognitive Science and Human Experience*. Cambridge, MA: The MIT Press.

Walker, Lawrence, Joe Walker, and Richard Hobbs, eds. 2007. *Linking Restoration and Ecological Succession*. New York: Springer.

Weiss, Paul A. 1973. *The Science of Life: The Living Systems—A System for Living*. New York: Futura Publishing Co.

West-Eberhard, Mary Jane. 2003. *Developmental Plasticity and Evolution*. New York: Oxford University Press.

Wolpert, Lewis, and Cheryll Tickle. 2011. *Principles of Development*. Oxford, New York: Oxford University Press.

Wood, Andrew J., G.J. Ackland, James G. Dyke, Hywel T.P. Williams, and Timothy M. Lenton. 2008. Daisyworld: A Review. *Reviews of Geophysics* 46(1): 1–23.

CHAPTER 3

Psychological Complementarity of Spectator and Actor

3.1 Indian Thought-Forms as Perceptual Gestalts

We began this study with a quote from Bohr, which suggests that the views of classical Indian and Chinese philosophers may illuminate epistemological issues in atomic physics associated with quantum theory. However, we have also seen that there is great resistance to forging dialogical links between science and Eastern traditions because it seems to open science to religious and mystical approaches to knowledge. While there is some justification for such concerns, it need not be taken to be a general rule that by paying attention to Eastern philosophies we necessarily embrace Eastern religious ideas or mystical claims. It is quite possible that by taking Bohr's recommendation seriously the epistemology of science could profit by incorporating Eastern philosophical insights without falling victim to either religious or mystical ideas. In this chapter we consider the case for opening science to a dialogue with Indian epistemological traditions.[1] The next chapter looks at the significance of dialogue with Chinese traditions.

In order to understand how complementarity connects with classical Indian epistemological notions, let us begin by looking at its implications for psychology. Consider Bohr's argument that extending the complementarity framework into psychology requires us to recognize that we are "spectators and actors in the great drama of existence." (Bohr 1958: 20) We have seen that the notion of gestalts allows us to understand how in the act of achieving a perceptual experience we function both as spectators

Arun Bala, *Complementarity Beyond Physics*,
DOI 10.1007/978-3-319-39784-9_3

and actors. The properties of a gestalt object are not merely given to us *qua* spectators confronting sensory stimuli, but are also constructed by us as actors who impose the interpretive framework through which these stimuli are read.

Moreover, there are good grounds to suspect that the gestalt structure of perception is rooted in the ubiquitous presence of grown properties in nature. We have seen that grown properties arise in crucial dependence on their environmental contexts through the position effect. Since humans have evolved in intimate association with a nature pervaded by grown properties, natural selection is likely to have promoted the evolution of our sensory apparatus to take account of environmental information when we set about identifying objects. In short, natural selection is likely to have made our senses take into account the position effect that produces many things in the world. Gestalt psychologists give precisely such an account of perception when they show that we identify objects not only by responding to stimuli from them, but also by taking into account stimuli from their surrounding contexts.[2]

However, the way in which we interpret stimuli in relation to the context within which they are embedded is not always biologically programmed—it could also be conditioned by experience and culture. Moreover, it often involves theoretical expectations about how things causally interact and interconnect in the world. Thus, for example, whether we perceive a circle of moving light as a band of fireflies, or the headlamp of a distant vehicle, depends on whether we think we are looking down a clearing in a forest or a long highway—that is, it depends on what we believe to be the environmental context of the stimuli we are receiving. Indeed the significance of theoretical beliefs and expectations in shaping our observations constitutes one of the major discoveries of recent epistemology of science, often formulated as the theory-ladenness, or theory-impregnation, of observations.[3] As a result, the role of genetically-programmed expectations in shaping percepts stressed by gestalt psychologists nearly a century ago has been widened to include the pervasive role of theoretical expectations in influencing how scientists experience the world. We experience gestalt percepts not simply because of the way our sensory apparatus takes into account contextual knowledge in ways that are biologically programmed, but also because it takes into account contextual knowledge in ways that are culturally shaped.

Many routes led to the discovery of the theory-impregnation of perception in modern philosophy and thought—experimental studies of perception that led to gestalt theories, anthropological studies that revealed the diversity of ways in which cultures organized perceptual experiences, philosophical analysis that decomposed perceptual experience into its component sensual and conceptual elements, and historical research that showed how scientific revolutions also brought about changes in the way scientists perceived, and not just conceived, the world. However, Indian thinkers were led to discover the theory-laden nature of perception much earlier when they argued that all perceptual experiences involve 'thought-forms.' By the expression 'thought-forms' they referred to the notion that our perceptual experiences do not depend on sensory stimuli alone, but also the conceptual expectations that we bring to bear on interpreting them.

That Indian epistemology should have made a discovery that came with great difficulty in modern thought may seem surprising until we examine how they arrived at such a radical perspective so early in history. The answer is that they were led to it through their use of meditation techniques. Such techniques are designed to deconstruct the interpretive frameworks that mediate our everyday experience of the world.[4] They promote a plasticity of sensory experience that led Indian thinkers to intuitively recognize the way we conceptually construct our everyday perceptual experiences in response to sensory stimuli from the world. It also led them to see that in ordinary states of consciousness our conceptual frameworks are so deeply entrenched by habituation that we suffer the illusion that our perceptual experiences are uncontaminated by conceptual beliefs. But deconstructing our conceptual frameworks through meditation techniques would directly reveal in the resulting intuitive experience how responsive our percepts are to our conceptual expectations. In short, the practice of meditation reveals that the objects of our everyday perceptual experience are thought-forms, that is, gestalts.[5]

The notion that meditation induced intuitive experience involves the introspective apprehension of gestalts may evoke strong objections since it appears to violate standard claims by Indian thinkers that intuitive experience transcends language. For example, Buddhist philosophers generally emphasize that their intuitive experience is empty of being (*sunyata*) and, therefore, both indescribable and unrepresentable.[6] If they are right then an account of intuitive experience as apprehension of perceptual gestalts would be self-indicting—it would be trying to use gestalts to model the indescribable and represent the non-representable.[7]

However, we have seen that similar claims had been made in relation to the properties of atomic objects in quantum physics. It has been said that the properties of atomic objects transcend language, and that we can only describe them in classical language that requires mutually exclusive concepts. But we found that treating quantum properties as growing in experimental contexts not only tells us about their fundamental structure, but also explains why the description of micro-objects appears to transcend language. Indeed language seemed limited only because the mathematical formalism of quantum systems does not represent them in terms of properties they actually possess, but in terms of properties they can grow in various mutually exclusive experimental environments. Although this could be taken to imply, in a sense, that the quantum system transcends the language used to describe it—at least to the extent that the properties used to represent it are not properties it actually has, but only properties it can potentially grow under different experimental contexts—it cannot be said that the system is not representable in language.

A similar problem concerning the limits of language can also be seen when we try to identify cells in a growing embryo. For example, whether we label a cell as an 'eye cell' or a 'kidney cell' in a growing frog embryo depends on its location. This follows because the cell is not named on the basis of its intrinsic properties or molecular structure, but in terms of what properties it will acquire by virtue of its location. The language that identifies the cell by its location does not represent the cell in terms of its structure or properties, but in terms of the structure and properties it will acquire in its current location. We are describing the cell not in molecular terms but functional terms—its role in its context. Even if this is deemed a limitation of language, since it does not label the cell in terms of what it is but what it will functionally become, it is nevertheless a description of the cell in terms of properties it will grow in its context. It does not require us to see cell properties as transcending language.

Hence, we cannot assume, even if Indian philosophers have generally claimed that ordinary language cannot adequately represent intuitive experience, that it is necessarily impossible to understand the nature of such experience. As with properties ascribed to quantum entities, or cells in a developing embryo, there may be good grounds for saying that the objects of intuitive experience transcend language. But this does not preclude understanding such an experience through the use of language, as we understand quantum objects and growing cells. Indeed, we will find that such experiences can be understood as the experience of gestalts, and that treating intuitive experience in this way can even explain why they are often seen as transcending language. Consequently we cannot take for granted the claims of those who achieve intuitive experience through meditation that it cannot, in principle, be understood through ordinary experience. Their judgment on this matter may be more a function of the psychological nature of this experience than any inherent logical grounding it possesses. Caught up in the revelatory experience induced by meditative practice they may find their intuitive experience so overwhelming as to suppose that it involves a qualitative break with ordinary experience. Nevertheless, they may be wrong.

Moreover, appeals to personal authority based on intuitive insights are questionable because even within Indian traditions there is controversy about the nature of intuitive experience. Buddhist, Hindu, and Jain philosophers all use similar techniques for achieving intuitive states of consciousness, but interpret their experiences quite differently, so that their appeal to the authority of intuition becomes suspect. Hence, nothing need preclude us from constructing a model of their intuitive experience provided it gives us insight into why Indian philosophers make the claims they do, or why they consider the objects of intuitive experience to transcend language. But whether Indian philosophers are justified in claiming intuitive experience to transcend all attempts to model it cannot be settled simply by appeal to the authority of their experience.

Even in psychology there is historical precedence for such an approach. Freud, for example, confronted the problem of giving an account of psychopathological behavior and experience, apparently incommensurate with normal experience, and not ordinarily accessible. But he solved the problem of showing us the logic of such behavior and experience by revealing it to be the result of the amplification of traits and defense mechanisms found in all of us (Freud 1949). He was able to do this because he exploited the continuity of the abnormal experiences he studied with certain normal experiences. This made it possible for him to show how we can penetrate the extraordinary structure and phenomenology of the psychopathological universe without requiring us to enter it directly. Similarly, if we can find a way to interpret the perceptual experiences in meditatively induced altered states of consciousness as accentuated developments of experiences in our normal states, then their logical structure would become explicable.

Moreover, such an approach seems plausible if we take the theory of biological evolution seriously. We would then expect meditation-induced intuitive experience to have some continuity with, and possess some affinity to, experience in everyday life—even if it is the result of an amplification of features in ordinary experience. (Happold 1984: 33–34) Indeed an evolutionary perspective would suggest that the intuitive experience cannot be rooted in faculties arising *de novo*, but must involve, and be an expression of, powers already of functional relevance to human beings in dealing with problems in everyday life. Otherwise it is difficult to imagine how such faculties and powers to generate the experience would have evolved through natural selection over human biological history. This suggests that those who achieve intuitive experiences must be deploying ordinary powers of perception—even if they have come to develop them beyond the normal range through intense discipline and training.[8]

Indeed support for the notion that Indian thinkers are responding to gestalts in their intuitive experience comes from a number of directions. First, consider the observation made by Schumacher and Anderson that gestalt images provide analogies to explicate the important distinction Indian Madhyamika philosophers in the Buddhist tradition make between the absolute undivided wholeness they perceive in intuitive experience and the multiplicity of empirical reality. They write:

> A crude analogy can be constructed if we think of Wittgenstein's figure of the duck-rabbit. The figure itself is neither a duck nor a rabbit; either way at the level of the figure nothing new arises. Obviously, the figure itself could not represent undivided wholeness, yet with this analogy we can say that the experience projects division out of undivided wholeness, and without such projections we have nothing but that undivided wholeness. (Schumacher and Anderson 1979: 82–83)

Actually there are grounds for thinking that Schumacher and Anderson may have noticed something more than a crude analogy, because in the altered states of consciousness induced by meditation practices, Indian philosophers may actually be discovering that all percepts have a gestalt structure.

Such a conclusion also accords with experimental evidence based on studying the effects of meditative practice, which suggests that it produces perceptual experiences of a gestalt character. Consider the study conducted by the psychologist Arthur Deikman on the effects of regular meditation upon perceptual experience. In his experiment Deikman made subjects meditate on a blue vase placed some distance from them. He instructed them specifically to "exclude all other thoughts or feelings or sounds or body sensations. Do not let them distract you but keep them out so that you can concentrate all your attention, all your awareness on the vase itself. Let the perception of the vase fill your entire mind."[9] In short he asked them to perceive the vase without engaging in any conceptual thinking or analysis by suppressing their thought processes.

Deikman obtained intriguing results from his experiments. Most subjects experienced the color of the vase to become more intense, and its shape to become more fluid and unstable over time. A number reported losing the sense of being separate from the vase—there was an experience of "merging" with the vase along with a loss of the consciousness of their body identity. Most revealing is the report of a subject who, asked to look out of the window immediately after a meditation session, reported:

> The view didn't organize itself in any way. For a long time it resisted my attempt to organize it so that I could talk about it.[10]

Deikman argues that the experiences of his subjects were the direct outcome of the deautomatization of conceptual structures that normally make possible the visual organization of our everyday perceptual experiences.[11] His interpretation suggests that one of the characteristic results of meditation practice may be the inability to describe in language the experience achieved because of the plasticity of the perceptual experience it generates in the altered state of consciousness it induces. The stability of normal experience is the result of the imposition of some single conceptual frame of reference to interpret incoming visual stimuli. The instruction to stop thought processes seems to suspend these conceptual organizing categories, thereby making perceptual experience malleable, and therefore resistant to a fixed linguistic description. This is precisely the sort of perceptual instability, or plasticity, we find with gestalt figures even in normal states of consciousness.

The hypothesis that Indian philosophers are responding to the plasticity of perceptual gestalts in their meditation-induced intuitive experiences explains a great deal of their epistemological claims. Many Indian philosophies maintain that the objects that arise in ordinary experience are, without our conscious recognition, conceptually elaborated forms—what they term 'thought-forms'. Indeed all Indian philosophies responsive to meditation-induced experiences make a distinction between experience of thought-forms (*samvrta* or empirical reality) and experience purged of thought-forms (*nirvana* or ultimate reality). This has inspired some modern writers to interpret Indian views along Kantian lines. Thus,

David Loy argues that the experience purified of thought-forms within the Indian tradition can be identified with the experience Kant calls the sensory manifold:

> One of the main ways Indian philosophy acknowledges the role of conception in perception is by making a distinction between savikalpa and nirvikalpa perception. Our usual perception is *sa-vikalpa* (with thought-construction), but there is the possibility of *nir-vikalpa* perception, which is "without thought construction" because the bare sensation is distinguished from all thought about it. The basis of both Sanskrit terms is *vikalpa*, a compound from the prefix *vi* (discrimination or bifurcation) and the root *kalpana* (to construct mentally). (Loy 1988: 42–43)

Loy also argues that perception without thought construction (*nirvikalpa*) is a variant of phenomenalism, though not one that can be identified with modern phenomenalism. Modern phenomenalism, according to him, does not question the ontological status of the subject though it concurs with the Indian view in rejecting any ontological status for the object. In Indian philosophy, and more specifically in the Buddhist philosophical tradition, phenomenalism is more radical in rejecting the substantiality of both subject and object.[12] Another writer Chandrahar Sharma was also earlier led to a similar phenomenalist interpretation of perception without thought construction (*nirvikalpa*):

> *Nirvikalpa* perception is the immediate apprehension, the bare awareness, the direct sense-experience which is undifferentiated and non-relational and is free from assimilation, discrimination, analysis and synthesis. (Sharma 1976: 194)

Furthermore, Loy argues that it is thought construction (*savikalpa*) made upon experience without thought construction (*nirvikalpa*) which produces a differentiated world of objects in empirical experience. Consequently, Loy argues, we have to consider that the general Indian tradition views empirical objects as really complex organized patterns of sensations, that is, as thought-forms or gestalts.

This interpretation follows the views of the renowned Indian logician Dignaga (c. 480–540), and his famous follower Dharmakirti (about 635CE).[13] Dignaga maintains that cognition has two phases—one in which it exists in the form of sensation, which is the first moment of cognition, and another which follows it when conceptualization intervenes so that it ceases to be pure sensation. This leads Dignaga to argue that our cognition of empirical objects in everyday experience is not a direct perception of them, but involves conceptualizing thought-forms out of the raw material of sensation. Dharmakirti went further to maintain that even the verbalizing of experience would not be possible without a prior conceptualization of sensual experience. He considered that sensation by itself cannot be verbalized—it is beyond language (*pratibhasa*). But he allowed that it is possible to recognize it in awareness without expressing it in language. (Bhatt 2000: 422–425)

Thus, it appears reasonable to interpret meditation-induced intuitive experience of empirical objects as the perception of plastic gestalts, or thought-forms in Indian terminology. In the intuitive experience of thought-forms there are sensations, which are apprehended independent of theory; and there are the objects of perceptual experience resulting from further conceptual elaboration upon this body of sensations. Thus, there are two levels of experience—first, the sensory manifold that is independent of conception (analogous to the bare configuration of a gestalt), and second, the perceptual experiences of thought-forms (gestalts) that arise on the sensory basis as a result of conceptual mediation.[14] We will find that it is precisely this two-leveled, that is, duomorphic, structure of the perceptual experience of thought-forms that makes Indian thought yield epistemological parallels to the complementarity viewpoint developed by Bohr to interpret quantum physics.

3.2 Thought-Forms and Indian Epistemological Traditions

However, before we look at the epistemological parallels to complementarity in Indian thought, let us see why interpreting meditation-induced intuitive experience as the experience of gestalts (thought-forms) allows us to make sense of the epistemological views of Indian schools of thought. Given the numerous schools of Indian thought we will approach our psychological hermeneutics of Indian epistemologies by first looking at the views of the second century philosopher Nagarjuna—the founder of the Madhyamika Buddhist School of philosophy. There are a number of reasons why Nagarjuna is a particularly apt choice for comparative study of classical Indian and complementarity epistemological traditions. First, many physicists and historians, including Bohr, have seen Buddhist epistemology in particular, among various Indian traditions, as exhibiting parallels with complementarity epistemology.[15] Second, Nagarjuna is historically an extremely influential thinker—his views generated intense controversies not only among Indian schools of philosophy, but also in Chinese and Japanese schools of thought. His far-reaching significance for Asian traditions of philosophy can be gauged by the fact that two historically seminal thinkers, Shankara and Zhu Xi, who founded the influential schools of Hindu Advaita philosophy in India and neo-Confucian philosophy in China, both developed their ideas in reaction to the Madhyamika. Indeed Buddhist influence on these schools was so pervasive that Shankara and Zhu Xi were both charged with being crypto-Buddhist thinkers in India and China respectively.[16]

In recent years his views have also become prominent in Western scholarship. The Buddhist scholar Seyfort Ruegg writes:

> Over the past half-century the doctrine of the Madhyamaka school, and in particular that of Nagarjuna has been variously described as nihilism, monism, irrationalism, misology, agnosticism, scepticism, criticism, dialectic, mysticism, acosmism, absolutism, relativism, nominalism, and linguistic analysis with therapeutic value. (Ruegg 1981: 2)

This has been endorsed by scholars of Buddhist philosophy Geshe Ngawang Samten and Jay Garfield who argue that Nagarjuna has been read by both modern and classical interpreters as an idealist, a nihilist, a skeptic, a pragmatist, and both a defender and critic of logic, and even a mystic. They conclude that these interpretations "reflect almost as much about the viewpoints of the scholars involved as do they reflect the content of Nagarjuna's concepts." (Samten and Garfield 2006: xx).

These multiple interpretations of Nagarjuna are reminiscent of the interpretation of Bohr as realist, positivist, idealist, instrumentalist, and even opportunist as we found noted by Mara Beller. Most significantly, as with Bohr, he is seen as paradoxical because his claims seem absurd and incoherent from the point of view of everyday experience. In particular, Nagarjuna's views, associated with many distinctive Madhyamika epistemological doctrines, such as the dependent origination of things, the emptiness of things, the identity of empirical and ultimate reality, the dichotomy of relative and absolute truth, and the logic of the negative tetralemma, seem to violate our standard conceptions of rationality and our everyday experience of the world. However, we will now demonstrate that many of the paradoxical claims of Madhyamika philosophy can be understood if they are seen as responses to perceptual gestalts in meditation-induced intuitive experience. This would serve to strengthen even further our argument that meditation-induced intuitive experiences involve the perception of gestalts.

Let us begin with the doctrine of emptiness—one of the most characteristic claims of Nagarjuna's philosophy, and central to many of his other claims. It is generally taken to assert the absence of being in things—also expressed as the 'relativity of things', the 'conditionedness of things', or the 'insubstantiality of things'. Nagarjuna traces his denial of being in things to the fact that they are dependent on our conceptual expectations. Hence, they originate by virtue of a synthesizing activity of the mind—what he terms the dependent origination of things in empirical experience. This implies the relativity, conditionedness, or insubstantiality of things. Consequently, for Nagarjuna, the things observed in the world are not self-subsistent, and therefore can be said to be empty of being (Ramanan 1978: 40–43).

Nagarjuna's doctrine of dependent origination and emptiness of being can be explained if we treat him as responding to gestalt objects in meditation-induced intuitive experience. Take the gestalt figure of the duck–rabbit configuration given earlier (see page 39). When we read the figure as a duck we conceptually organize the elements of the configuration in a specific fashion—otherwise we would not see the duck. To someone who fails to see the duck we tend to offer an intellectual interpretation along lines such as "This is its neck, that its beak" and so on. In short, we attempt to guide the person toward making a thought-mediated synthesis of the elements in the indeterminate configuration to achieve the experience of seeing a duck. It follows that the duck the person sees has no independent existence, but arises in dependence upon the conceptual context through which it is viewed. Hence, it has no being-in-itself apart from its interpretive context, and can be said to arise in dependent origination upon this context. Since all objects in the world are experienced as gestalts in the plastic altered states of consciousness induced by meditation, we can say that Nagarjuna is claiming that all objects in the world are subject to dependent origination and, therefore empty of being.

There can be an objection that Nagarjuna is glossing over a significant difference between claims we make about our experience of objects in the world, and the world of objects-in-themselves. After all the determinate duck or rabbit thought-forms that arise out of the indeterminate configuration are not things in the physical world but only images we experience. In short, they are merely thought-forms. But Nagarjuna appears to have projected his experience of thought-forms in intuitive experience into a doctrine about the nature of things in the world, by claiming that things in the world have no being-in-themselves. Thought-forms have no being-in-themselves because they arise in dependence on a synthesizing activity of the mind, but this cannot be said of physical objects in the external world.[17] These have being-in-themselves by virtue of existing independent of us, even if our cognition of them is made through thought-forms that have no being-in-themselves.

However, such an objection is not germane for our purposes of developing a hermeneutics of Nagarjuna's specific epistemological doctrines. After all Nagarjuna does not recognize a substantial world behind the world of appearances and, in this respect, he is faithful to the general Buddhist philosophical rejection of any substance either as an objective ground outside (in the form of physical objects, say) or subjective ground within (in the form of souls or egos, say). He appears concerned with nothing more than an analysis of thought-forms or gestalts, which are for him the only things in the world. Unlike physical objects, or egos, thought-forms are appearances, which are empty of being and arise in dependence upon our conceptual interpretations.

Let us for the purpose of our hermeneutic analysis of his views, therefore, follow him by taking the intuitive experience of thought-forms induced by meditation as revealing what the world is really like. This would allow us to appreciate why he is led to conclude that everyday experience has reified thought-forms into substantial objects so that we are unable to recognize their conceptually mediated nature. Nagarjuna maintains that through meditative deconstruction we can achieve the intuitive knowledge (*prajna*) that would free us from being enmeshed in these intellectual constructions, by enabling us to experience immediately the sensory manifold uncontaminated by the constitutive intellect (*buddhi*). He suggests that when we leave our intuitive state of awareness in which all things are apprehended as plastic thought-forms, and re-enter the world of everyday experience, by taking on the interpretative intellectual functions of the mind, we would experience everyday objects as crystallizing out of the sensory manifold.[18] This sedimentation of things and entities out of the sensory manifold in interdependence upon each other, and as a result of the synthesizing activity of the mind, would give us direct intuition of not only the dependent origination, but also the emptiness, of all things.

Gestalt figures allow us to appreciate Nagarjuna's claims without leaving our ordinary state of consciousness. The duck–rabbit figure, for example, can be reified so that we can only see, say, either the duck or the rabbit but not both. Or we can free ourselves from intellectual construction so that only the configuration is experienced. Reintroducing conceptual mediation we can perceive the duck (or rabbit) to sediment out of the configuration. Thus, our experience of the gestalt figure serves as a model for understanding what happens when someone moves from deautomatized experience, when all things are intuitively experienced as having a gestalt structure, to everyday experience when the imposition of singular conceptual frameworks reifies them into crystallized objects. It also enables us to understand why the duck and the rabbit are thought-forms empty of being since they originate in dependence on conceptual interpretations.

It is also important to note that Nagarjuna does not view the emptiness doctrine as solely designed to reveal that objects in the world are only thought-forms—he also intends such knowledge to lead us to suspend our conceptual constructions so that we would apprehend a reality independent of all conceptual projections. This is evident in Nagarjuna's opening dedicatory verse in his seminal work *Mulamadhyamakakarika*.[19] The comparative philosopher Kenneth Inada describes this verse as embodying the Madhyamika creed because it "expresses the whole philosophy of the Madhyamika in a nutshell". Nagarjuna proclaims the goal of his work in this dedicatory verse:

> I Pay Homage to the Fully Awakened One,
> The Supreme Teacher who has taught
> The doctrine of relational origination,
> The blissful cessation of all phenomenal thought constructions.
>
> (Inada 1970: 38–39)

In recommending liberation from all conceptual entrapment, that is, thought-forms, as the illuminative project of the Madhyamika, Nagarjuna is led to make a distinction between two kinds of reality—an unconditioned reality, experienced after we have freed ourselves from conceptual projections, and a conditioned reality, enmeshed in reified thought-forms that constitute our everyday world of objects. According to Nagarjuna, the conditioned reality is only how the unconditioned reality appears after conceptual interpretations are imposed upon it. This leads Nagarjuna to caution us against taking the conditioned and unconditioned realities to be different and separable simply because they are distinguishable. Doing so would mislead us since the conditioned reality is always dependent upon the unconditioned, although the unconditioned reality can exist independent of the conditioned reality. Nagarjuna's doctrine of the distinction between, and the inseparability of, two levels of reality is stressed by Ramanan in his study of Nagarjuna's philosophy:

> It is the basic conception in the philosophy of Nagarjuna that while the indeterminate [unconditioned] reality is the ground of the determinate [conditioned] entities, it is only the ultimate nature of the latter themselves and not an entity apart from them. (Ramanan 1978: 39)

This distinction between two levels of reality with one dependent on the other, but not vice versa, can also be explained if we treat Nagarjuna as responding to the experience of all things in the world as thought-forms or gestalts. For if we take the configuration of the duck–rabbit figure as the unconditioned reality, and its appearance as either the duck or rabbit thought-form when we superimpose conceptual categories upon it as the conditioned reality, it follows that the unconditioned and indeterminate configuration is the ground of the conditioned and determinate thought-forms. Although distinguishable from both duck and rabbit thought-forms the unconditioned configuration is not an entity apart from them, but that upon which they crucially subsist. Indeed referring to the duck-rabbit gestalt the following propositions would hold true:

1. The configuration out of which the duck and rabbit thought-forms sediment is the unconditioned reality that is not a thought construct. It is given as an ultimate to which our orientation can only be one of pure receptivity as spectators.
2. The duck and rabbit thought-forms are conditioned objects. They constitute empirically perceived entities that have been made to crystallize out of the unconditioned ground of the configuration. We do not relate to these thought-forms purely as spectators because we are also involved in constituting them as actors.

3. The distinction between the configuration and the duck (or rabbit) is not a reason for treating the duck (or rabbit) as an entity distinct from the configuration. The configuration is the unconditioned ground of the duck (or rabbit) thought-form, and not an entity apart from it.

Nagarjuna's conception of two levels of reality leads him to an associated notion of two levels of truth. It makes him distinguish what he terms empirical truth (*samvrti satya*) and absolute truth (*paramartha satya*). Empirical truth is associated with knowledge claims based on our ordinary experience of the world of conceptually mediated objects as given independent of thought. Absolute truth is associated with knowledge claims based upon intuitive recognition of the unconditioned reality upon which they are projected as thought-forms. He also maintains that empirical and absolute truth should not be treated as involving assertions about distinct classes of objects. Nevertheless Nagarjuna considers absolute truth to have a higher epistemic stature than empirical truth. This leads him to maintain that when we speak of empirical truth we are only using the term "truth" by way of concession to the conventional language of everyday experience. (Murti 1960: 251) In an ultimate sense, however, absolute truth is the only truth.[20]

Adopting a gestalt view of intuitive experiences enables us to make sense of Nagarjuna's notion of two levels of truth. When we respond to the gestalt figure we could either make claims about the configuration itself, or about the thought-forms elaborated upon it. What we say about the configuration is about something really out there independent of our beliefs, and these claims can be treated as absolute truths about some non-constructed entity. However, when we make claims such as "This is a duck," or "This is a rabbit," we are referring only to thought-mediated forms. These claims can be said to be true only in concession to ordinary linguistic usage, but they are not true of any mind-independent object. Moreover, all claims about thought-forms do not refer to anything distinct from the configuration. It is possible to argue that absolute truth is about the unconditioned configuration apart from the thought-forms projected upon it, but empirically true claims are only about conditioned thought-forms crucially dependent on the configuration and interpretations we bring to bear upon it. The former has a higher epistemic stature than the latter—or so we would claim if we went along with Nagarjuna's notion of two levels of truth.

Assuming the gestalt structure of intuitive experiences can also illuminate the apparently paradoxical structure of Nagarjuna's famous Madhyamika *catuskoti*, or negative tetralemma. (Murti 1960: 38–40)[21] The *catuskoti* can be expressed very simply—it essentially asserts that the structure of the world is such that, given a proposition P about the world, we may be able to *deny* all of the following:

P, not-P, both P and not-P, neither P nor not-P

Thus, the claims of the tetralemma seem to violate the laws of standard logic such as the law of noncontradiction, and the law of the excluded middle.[22]

The Madhyamika tetralemma has intrigued and irritated philosophers ever since it was formulated. Its apparent assault on logic has engendered numerous interpretations designed to make sense of it. It has been generally assumed that Nagarjuna uses the tetralemma to expose the antinomian character of reason, or the limits of logic, in handling intuitive experience. One problem with treating the *catuskoti* in this manner is that Nagarjuna himself appeals to reason and the laws of logic, whose limits he is supposedly exposing, when he develops his critique of other philosophical positions.[23]

However, the gestalt view of intuitive experience accounts for the *catuskoti* without requiring us to see him as adopting the self-refuting strategy of using reason and logic to repudiate reason and logic. Take the duck-rabbit figure discussed earlier. Let the proposition P be "It is a duck". In the conceptual context where we see the configuration as a duck, the proposition P has to be affirmed. In the conceptual context where only the rabbit is seen, the proposition not-P would have to be asserted. In the conceptual context where both of the images are recognized as perceivable we would have to affirm the conjunction. Such a conceptual context is actually a metaconceptual standpoint that includes both the duck and rabbit conceptual contexts. In the context where we oscillate between both the duck and rabbit conceptual contexts, so that when in the duck context we have to deny that it is a rabbit (and vice versa), the disjunction has to be affirmed.

The *catuskoti*, however, is not proposed as an affirmation but as a denial of P, not-P, both, and neither. (Murti 1960: 129–132) Hence, it is a denial of all propositions based upon conceptually structured contexts—whether it is a single conceptual context that involves only the P-context or only the not-P context; a metaconceptual context that includes both the P and not-P contexts; or an alternating context that shifts from the P-context to the not-P context. Each of these positions involves perceiving through conceptual interpretations; all of them fail to lead us to the indeterminate configuration upon which these percepts are erected. All of them entrap us in thought-forms that do not reveal the unconditioned configurative ground upon which they are erected. Hence we have to deny all of them: P, not-P, both and neither.

Thus, by adopting the gestalt view of intuitive experience we can account for the negative tetralemma without assuming that Nagarjuna is calling for the repeal of the laws of reason or logic. The tetralemma is not a critique of logic but our understanding of the nature of perceived objects in the world. It is Nagarjuna's attempt to point to the dependence of the objects of empirical experience on our cognitive contexts. It is deployed by him as a heuristic to point to the need to transcend all conceptual models rather than express reservations about reason or logic. Such reservations would subvert even his position, since he too deploys standard reason and logic to both develop his views and criticize his opponents.

Nagarjuna maintains that in everyday empirical experience we not only fail to perceive the indeterminate ground of our percepts, but are also attached to the construction of percepts from a single conceptual standpoint—a standpoint we are socialized into adopting as the only possible one. Reading the negative tetralemma from a single conceptual context of empirical experience would lead us into logical contradictions. To appreciate this point consider what would happen if we were able only to perceive the configuration as a duck but not as a rabbit. Attempting to interpret the propositions of the tetralemma from this one conceptual context we would see P and not-P as logically incompatible propositions—as contradictory—and not contrary ways of seeing the same configuration. The situation would appear even more paradoxical with the propositions "Both P and not-P," and "Neither P nor not-P," for then we seem to be violating both the law of non-contradiction and the law of the excluded middle.

Hence, the Madhyamika tetralemma is not designed to point to the limits of reason or logic—it is really intended to point to the limits of empirical experience which, being generally confined to a single conceptual context, leaves us unable to recognize the contextual relativity of the empirical objects we perceive; neither does it enable us to isolate the sensory manifold out of which they arise (except through careful philosophical analysis). It is only when we achieve intuitive knowledge of this indeterminate manifold, by using meditative practices to suspend our usual thought-constructions, that we recognize the logical coherence of the *catuskoti*. It is at that point that we come to see directly that what the *catuskoti* subverts is not standard logic, but our normal assumption that the established thought-forms that define our experiences are the only ones possible. The negative tetralemma suggests that perceptual judgments are context dependent, and what appears to violate logic seen through everyday perceptual experience, is indeed what we are logically compelled to conclude if we achieve the richer intuitive experience of thought-forms that meditation renders possible.

The gestalt interpretation can also be extended beyond the Buddhist Madhyamika tradition to explain epistemological positions adopted by some Jain and Hindu schools of Indian philosophy that also appeal to meditation-induced intuitive experience. These schools can be seen as making epistemological responses, albeit different from the Buddhist philosophical tradition, to intuitive experience of thought-forms. Consider the sevenfold perspectivism of Jain philosophy—what is often called the *syadvada* doctrine closely associated with the non-onesidedness-doctrine (*anekandavada*).[24] The non-onesidedness-doctrine is often set in opposition to the four-fold *catuskoti* of the Buddhist philosophers. But if the experience induced by Jain meditation practice, which is largely similar to Buddhist practice, also has a gestalt structure, then we can explain Jain perspectivism as another response to gestalts.

To appreciate this point let us look at the Jain non-onesidedness doctrine more closely. According to comparative philosopher Ninian Smart this doctrine maintains that regarding any claim we make about the world it is possible to assert the following seven judgments: (1) 'Maybe it is'; (2) 'Maybe it is not'; (3) 'Maybe it is and it is not'; (4) 'Maybe it is inexpressible'; (5) 'Maybe it is and is inexpressible'; (6) 'Maybe it is not and is inexpressible'; and (7) 'Maybe it is and it is not and is inexpressible'. Smart interprets the Jain position along the following lines:

> I may assert that a temple flower exists, but in making a distinction I am saying that a tiger does not exist at that spot. But there is no way of expressing how it is both a temple flower and not a tiger. And so on. Not only does this Jain scheme place judgments under seven forms but it prefaces each with a 'Maybe'; *syad* or 'Could be', from which the doctrine is called *syadvada* or 'Could-be-ism'. This sense of the perhaps reflects Jainism's thought that all viewpoints are partial. This is illustrated by the simile of the blind men

grasping different parts of an elephant and giving different reports—all true up to a point but all false too. (Smart 2000: 26–27)

However, being ignorant of the fact that the Jain philosophers are actually referring to their experience of plastic gestalts in meditatively induced experience, Smart completely misses the point of the Jain maybe (*syadvada*) doctrine. In the first place he does not explain why the Jains had to add a 'maybe' to the claim that a temple flower exists—this would arise only if I am also led to believe that it may not be a temple flower. Second, Smart's view that there is no way of expressing how something is both a temple flower and not a tiger is unconvincing—after all he has expressed the claim in the process of telling us what the Jains considered inexpressible! Finally, the example of the blind man with the elephants fails to explain why the Jain thinkers considered there to be only seven viewpoints, and not some number more or less than seven.

However, if we take the Jain philosophers to be referring to the experience of plastic gestalts then their position becomes quite reasonable. Take the duck-rabbit figure again. Suppose we now formulate the claim to be "It is a duck", and take the term 'inexpressible' to refer to the configuration since it cannot be described in words. Then we could assert the following: (1) 'Maybe it is a duck'; (2) 'Maybe it is not a duck'; (3) 'Maybe it is and it is not a duck'; (4) 'Maybe it is the inexpressible configuration'; (5) 'Maybe it is a duck and is the inexpressible configuration'; (6) 'Maybe it is not a duck and is the inexpressible configuration'; and (7) 'Maybe it is a duck and it is not a duck and is the inexpressible configuration.' Moreover *these seven possibilities exhaust all possible claims we could make.* Since we can assert or deny "It is a duck," assert or deny "It is not a duck," and assert or not assert "It is the inexpressible configuration," there are eight possible ways of combining these claims. However, the assertion "It is not a duck and it is a duck" merely repeats the assertion "It is a duck and it is not a duck." Hence, there are only seven possibilities in the Jain septalemma.

Since both the Buddhist tetralemma and the Jain septalemma can be seen as responses to perceptual gestalts it lends credence to the claim that meditation-induced gestalt experiences motivate the epistemological positions of Indian thinkers. Though Jain thought formulated its epistemological implications differently from the Madhyamika, it is nevertheless possible to see both schools as motivated by experiences not accessible to ordinary states of consciousness, but which arise when we deautomatize the conceptual categories mediating perception by meditation techniques.

It is also possible to see Hindu Advaita epistemology as responding to perceptual gestalts in intuitive experience. According to the comparative philosopher Chakarvarthy Ram-Prasad:

> [T]he Advaitin says that the world is as the snake to Brahman's rope. The real point is not that the world is an illusion; this very example, the Advaitin is aware, requires a distinction between illusion and object in the world. It is, rather, that just because we had an experience of something we should not conclude that the something exists just as we experienced it. We should not rule out the possibility that something is not what we took it to be. The world is "real" enough, in terms of our experience of it—in terms of our seeing and touching and moving and thinking about it. But from such experience we cannot conclude that it is ultimate. Advaitins use special terms to explain the world as penultimate; it is only empirically or provisionally real; it is phenomenal (capable of being experienced); it is indeterminate between the real and the unreal. (Ram-Prasad 2005: 39–40)

The hypothesis that Advaitins are responding to gestalts can also make sense of their epistemological views described above. Consider the Advaita Vedanta claim: "The world is as snake to Brahman's rope." If we take the rope (Brahman) as the noumenal ground to be analogous to the duck-rabbit configuration, then the snake (world) as the phenomenal appearance would be the projection upon it, in the same way the rabbit or duck is projected upon the configuration.

Finally let us see whether appeal to gestalts can make sense of Nagarjuna's dialectical methodology. Unlike the tetralemma which is only intended to reveal the dependence on conceptual contexts of empirical objects in everyday experience, Nagarjuna's dialectics is designed to liberate us from all such conceptual contexts. According to Murti, there are three stages in the dialectic of Nagarjuna. (Murti 1960: 140–143)[25] In the first stage—the stage of dogmatism—philosophical systems are speculatively constructed. As a result we come to experience reality through thought-forms (*vikalpa*) mediated by such systems, and the resulting empirical presentations experienced (*samvrta*) are treated as independently given absolutes. However, since there are different philosophical systems each of which gives rise to different philosophical constructions that project different perceptual experiences, we are led into the arena of interminable conflicting systems of philosophy.

According to Murti, this leads to a second stage when dialectical critique shows the limits of all fixed views by raising awareness to a level where what appeared as independently given absolutes now appear as possibly illusory. As a consequence, he notes, there develops the realization of emptiness—the awareness that what we take as real may be an intellectually mediated illusory construct (Murti 1960: 140).

Murti maintains that this opens the door to the final stage when there arises absolute knowledge in intuition, brought about by suspending all thought constructions (*drstijnana*). The result is a revelation of something already present, but previously obscured by thought-forms (*samvrta*)—it is the knowledge of the unconditioned (*nirvikalpa*) that is not a construct. This is the intuition of absolute truth—a truth free from dependence on judgments responding to thought-constructions.

Murti argues that the goal of Nagarjuna's philosophy is to take us through the three stages of the dialectic—from dogmatism to criticism and, finally, to intuition. Nagarjuna's dialectics, according to Murti, is designed to deconstruct the conceptual foundations of all speculative philosophy so as to open the way for intuitive apprehension of reality unconditioned by thought-forms. Thus, the dialectic is not merely a philosophical method—it is also a soteriology designed to transform our sensibilities. It guides us along a path that leads us to gradually relinquish the conceptual matrix through which we come to know the objects of the empirical world. Thereby we come to cognize not only the emptiness of empirical objects, but also achieve a direct intuition of their unconditioned indeterminate ground.

In this way, according to Nagarjuna, the dialectic removes the veil of ignorance (*avidya*) that invests the unconditioned ground with false appearances. Ignorance performs two functions. In the first place, it is obscurative (*avarana*) because it veils the nature of things from us. Second, it is constructive because it creates an illusory appearance (*asatkhyapana*) through our propensity for conceptual construction (*sankalpa*) by making us see the unconditioned ground, which is indeterminate, as determinate because we view it through thought-forms.[26]

Nagarjuna's dialectics can also be understood by adopting the gestalt view of the objects of intuitive experience. The three stages of dogmatism, criticism, and intuition can then be seen as follows. In the first dogmatic stage a figure is perceived as a duck only, or a rabbit only, but not both. There is no transcendence of the single conceptual context adopted. Critical examination leads to the recognition of the alternative possibilities of seeing the figure as either a duck or a rabbit and, thereby, realizing the non-tenability of dogmatic adherence to any one way of seeing it. This opens the door for intuitive experience of the configuration unconditioned by conceptual construction out of which both the duck and rabbit have been constructed. It removes the original ignorance that invested the configuration with false appearances, that is, made it appear to be a duck or rabbit when it was neither. The ignorance (*avidya*) was not only obscurative in that it hid the configurating ground; it was also constructive because it projected the illusory appearance of a duck or a rabbit upon it. Thus Nagarjuna's dialectics can also be seen as involving a re-education of our perceptual sensibilities that goes beyond mere revision of our modes of conceptual thinking.

It is evident that Nagarjuna's epistemological views can be explained as a response to gestalts which arise in meditation-induced intuitive experience. However, there may be some reservations about such an interpretation that I would like to address. In the first place the identification of the configuration that grounds the gestalts with what Nagarjuna refers to as the unconditioned indeterminate may appear untenable. Nagarjuna claims the indeterminate is indescribable in language, but the configuration out of which the duck-rabbit figures arise can be described using language, albeit with some effort and difficulty. Hence, it may be said, although gestalts provide useful analogies for understanding the Madhyamika intuitive experience, the configurative ground of gestalts cannot be identified with Nagarjuna's indeterminate.

This objection deserves serious consideration. Indeed, although I have referred to the gestalt configuration as indeterminate this assertion has to be qualified. Relative to the determinacy of the duck or rabbit figures, the configuration is clearly indeterminate for only by imposing further conceptual structuring upon it do we obtain these images. However, on its own, and in isolation from them, it also has a well-defined determinate structure. Being determinate, the configuration is describable in language and cannot be identified with Nagarjuna's unconditioned that he views as radically indeterminate, and beyond any linguistic description.

The important point this objection highlights is that the configuration itself is also a conceptual construct. It is also a thought-form ultimately elaborated out of sensory stimulation falling on the retina. After all, the perceived configuration is the result of billions of photons impinging on the eye every second and, moreover, stimulating its retinal cells in a continually shifting pattern as the eye roves over the figure in the act of perception. Yet in spite of this torrential flux of retinal stimulation a stable image of the configuration is sustained. This is possible only because we read the flood of stimuli through conceptual organizing principles that create it as a stable and well-defined entity.[27]

If we assume that meditation-induced conceptual deautomatization can also be made to effect the suspension of principles that stabilize the configuration we experience, then even the configuration would become more unstable and fluid, and finally lose its identity. At that point it would become completely indescribable. Moreover, we could also reverse the process. Starting from the deautomatized state, we could set out to re-impose by stages the suspended conceptual categories, so that at first the configuration, and then the duck or rabbit figure, is made to crystallize out of the indeterminate ground. These determinate entities would then appear to arise out of the indeterminate sensory manifold. Clearly, the process described allows us to suppose that the sensory manifold is separated from the determinate figure (such as the rabbit) by many levels of conceptual elaboration all of which may be deautomatized. Moreover, each stage of this series of stages of conceptual elaboration and consolidation would appear determinate relative to its preceding stage and indeterminate relative to its successor stage.

Hence, the relative determinacy of the configuration compared to the Madhyamika absolute indeterminate does not render untenable our attempt to offer a psychological hermeneutics of Nagarjuna's doctrines. What is important for comprehending these doctrines is the duomorphic structure of all perceptual experiences that arise as thought-forms.[28] It is the indeterminacy of the configuration relative to any thought-form which arises upon it, and not the absolute indeterminacy of such a configuration, which is needed to explain Nagarjuna's doctrines of dependent origination, emptiness, two-fold truth, two-fold reality, as well as his tetralemma and dialectical method. The additional thesis of an absolutely indeterminate basis for thought-forms is only required to support Nagarjuna's claim that the process of conceptual deautomatization can lead us to the awareness of an ultimate that transcends language altogether.

Clearly the hypothesis that Indian thinkers are responding to the gestalt structure of objects in intuitive experience explains in an elegant and economical fashion many epistemological views of Buddhist, Jain, and Hindu schools of philosophy. Nevertheless, it may be felt that the gestalt model offers an unusually reductive account of intuitive experience by failing to address the broader ethical, spiritual, and soteriological issues raised by

Indian thinkers. With some justification, critics of the above approach could contend that no model of perceptual gestalts could ever conceivably do justice to these more expansive philosophical claims that are integral to any comprehensive hermeneutics of both the intuitive experience achieved through meditation and diverse Indian epistemic views.

There are two responses I would like to make to this objection. To begin with, the model proposed is not intended to address the broader religious and spiritual claims of these Indian schools. It is solely concerned with explaining their experience of perceived empirical objects, and the doctrines that emanate thereof, in order to lay the basis for explaining the Bohr parallels between quantum physics and Eastern doctrines. The charge that the gestalt model is unable to explain the wider Eastern claims is irrelevant to the issue of whether their account of empirical experience emanates from responding to plastic gestalts in meditation-induced intuitive experience. Moreover, it is only reasonable to suppose that the radical conceptual deautomatization that transforms perceptual sensibilities also produces dispositional, affective, and spiritual mutations that lie beyond the domain of applicability of a perception-oriented gestalt model. But this limitation does not impugn the applicability of the gestalt model to intuitive perceptual experiences.

Nevertheless, even if we have to confine the scope of the model to accounts of perceived empirical objects in intuitive experience, it is noteworthy that most Buddhist, Jain and Hindu schools of philosophy, in contrast to the materialist Lokayata and Carvaka traditions that reject meditation-induced experiences as having noetic value, retain the general duomorphic distinction between an unconditioned reality and an empirical reality, and the notions of higher and lower truths for all areas of knowledge. This suggests that the gestalt model, while it cannot be indifferently extended to include non-perceptual experiences that arise from meditatively mediated deconstruction, can nevertheless serve as a powerful metaphor to illustrate the epistemological structure which Indian thinkers see as applicable to a much broader spectrum of experience.[29]

3.3 Madhyamika and Complementarity Epistemological Parallels

The gestalt structure of the objects of intuitive experience allows us to explain the strange parallels noted by Bohr between Buddhist doctrines and the epistemology of complementarity. We have seen that quantum properties grow in the experimental context selected to make observations. Similarly, gestalts, or thought-forms, also arise in the theoretical context selected to make observations. Both are cases illustrating the context dependent origination of observed properties. We will find that this explains why quantum physicists and Indian philosophers arrived at similar epistemological views despite confronting quite different objects in their respective domains of inquiry.[30]

To see this let us begin by considering the configuration that gives rise to the duck and rabbit images. By analogy with the wave function of quantum theory that 'collapses' to a specific observed value in the context of observation (such as a specific position or momentum), we can refer to it as a gestalt function because, it also 'collapses' to a specific value (say, duck or rabbit) in the theoretical context of observation. Prior to the establishment of the context of observation the gestalt function does not predict an unambiguous outcome for the result of observation, but merely constrains its possible outcomes. It is only within the context established by the conditions of observation that the configuration, like the wave function, "collapses" into a definite value, such as a duck or a rabbit.

This dependence of the observed result on the choice of observational context also allows us to develop gestalt parallels to the EPR situation. Consider once again the duck-rabbit figure. The objects marked X and Y in the figure can be seen as "beak" and "nape" when we choose the observational context E1 in which the gestalt function is interpreted as a duck; but in the different context E2 in which the function is read as a rabbit, X and Y become identified as "ear" and "throat" respectively. Clearly, the two contexts E1 and E2 are mutually exclusive. We cannot perceive X as "beak" and "ear" at the same time; nor can we see Y as simultaneously "nape" and "throat". Furthermore, there are correlations between our perceptions of X and Y. When X is perceived as a beak, Y is seen as a nape; when X is an ear, Y is a throat.

The correlations between our perceptions of X and Y are analogous to those in the EPR example. Suppose an observer records perceiving X as a beak. Then we can predict with certainty that Y would be perceived as a nape. Yet, there is no question of a mechanical disturbance emanating from X and influencing Y. It is the observational situation E1 in which the gestalt function is seen as a duck that effects the correlation. Similarly, a different observational situation E2, where the observer records perceiving X as "ear" would lead us to predict with certainty that Y would be seen as "throat".

Does this mean that Y is simultaneously a nape and a throat? Do both these descriptions have simultaneous reality? Clearly, this is not the case, since Y cannot be treated as being a nape or a throat until after observation has reduced the gestalt function to a duck or a rabbit. What is observed originates only within a context of observation.

Second, as in the quantum case, we cannot separate the properties of the system being observed from the observational arrangement we have selected. The choice of observational context crucially affects the property observed. For example, the property of X being such-and-such is determined by which observational arrangement we select; that is, whether we select the "duck" or "rabbit" arrangement. Our choice of the context of observation affects the results observed.

The notion of mutually exclusive observational arrangements, so closely linked to the doctrine of complementary properties in quantum physics, is also found in the gestalt situation. By choosing one or another of the observational contexts E1 and E2 we can determine Y as "nape" or "throat" correlated to the observation of X as "beak" or "ear". These two contexts are mutually exclusive and cannot be added together to support a single picture. But both are necessary to give a complete description of all possible results of observation. The notion of complementary properties—mutually exclusive but equally necessary—introduced by quantum physics has its parallels in gestalt perception.

Finally, Bohr's dual language view—namely, that the language of classical physics and the language of the mathematical formalism of quantum theory are both necessary—is also paralleled in the gestalt situation. In this case the language analogous to classical physics is the one used to describe the system in terms of percepts that we ordinarily experience using terms such as "beak", "ear", "throat", and so on. This is similar to the language that uses terms such as "position," "momentum," and so on. The other is the language analogous to the mathematical language of the wave function. It is used to describe the gestalt function, or configuration, in terms of properties it can acquire under different interpretations. This might involve a more abstract description of possibilities—for example, "If this is an ear, then that is a throat but if it is a beak then that is a nape." This is a much more abstract language of conditionals. It is analogous to the way the wave function describes the system in terms of possible results of measurement under different conditions of observation, such as determining its position or momentum. In both cases—the gestalt function and the wave function—a single, concrete, and consistent description is not possible. Neither is there any way of going beyond to another description in an improved language: the problem is not due to an inadequate linguistic system, but to the ambiguity inherent in the mutually exclusive possibilities in the gestalt function and the wave function. Thus the complementarity perspective can be deployed to interpret the gestalts to which Nagarjuna is responding.

Let us now see how Nagarjuna's epistemology can also be used to interpret quantum phenomena. Take Nagarjuna's notions of dependent origination and indeterminate ground. The doctrine of dependent origination (*pratityasamudpada*) can be applied to the wave and particle properties an atomic object exhibits. We can argue that the micro-entity acquires spin in one direction in one context, and spin in a different direction in another context. But since the spin properties did not pre-exist the context, they can be seen as originating in dependence on the context. Consequently we cannot project these observed spin properties that have dependent origination onto the micro-object prior to its entry into the experimental context. Hence, the micro-entity can be said, like the Madhyamika indeterminate ground, to be in an indeterminate state which becomes determinate only after it acquires a well-defined spin property in the context of its measurement.

However, there is one crucial difference in the way Nagarjuna and physicists construe empirically observed properties. For Nagarjuna such properties are empty of being because they are projected upon the ground as a result of conceptual construction. By contrast, physicists take observed quantum properties to be real, although they arise as a result of the interaction of the micro-object with the observing apparatus. The reason for these differences in judgment is the dissimilar roles played by the human agent in the two cases—in the case of gestalts, the agent freely selects the theoretical apparatus through which observations are made; in the case of quantum observations, the agent freely selects the experimental apparatus for making observations.

However, not all Buddhist philosophers subscribe to the Madhyamika interpretation of dependent origination as conceptual dependence of observed properties. The earlier Theravada tradition of Buddhist philosophers see the world as constituted by point events that originate and subside in causal dependence upon each other. (Murti 1960: 7) This is analogous to the notion that the world is atomic in structure, except that unlike modern atomism, which sees atoms as persisting over time, the Theravadins see every single atom as a point event that comes into existence for an instant in causal dependence on other instantaneous atomic point events in its environment. Hence, the

Theravadin causal notion of dependent origination that point events arise by virtue of their relations to other point events, is quite different from the Madhyamika conceptual notion of dependent origination that perceived objects arise from sensations by virtue of their relations to theoretical beliefs.

It may seem that the Theravada causal notion of dependent origination is more in consonance with the quantum viewpoint than the Madhyamika conceptual notion of dependent origination. By making origination dependent on objective events independent of our beliefs, the Theravada view does not adopt the notion of empirical reality as mind created. However, it still remains the case that the Theravada position does not conform to the physical realism of quantum theory, since it sees its elementary point events as instantaneous in time and atomic in space—they are not the permanent atoms of quantum physicists.

Moreover, the Theravada and Madhyamika views may not be quite as opposed to one another as normally understood. If an entity arises in causal dependence upon other entities around it, as the Theravada maintain, then we can use theoretical knowledge of such causal relations to exploit environmental information around the entity to facilitate its identification. This is precisely what lies at the root of the modern theory-ladenness or theory-impregnation of observations thesis. The causal version of dependent origination espoused by the Theravada tradition then becomes the basis for the conceptual interpretation of the Madhyamika. This makes it possible to see the Theravada causal and Madhyamika conceptual versions of dependent origination as interrelated and symbiotic rather than conflicting.

Such an integrative account would not assume, as Nagarjuna does, that the doctrine of emptiness (*sunyata*) implies that things have no being, or are empty of being. It would see it as simply stating that they have no being-in-themselves because their being arises in dependence upon other things. This is indeed the Theravada viewpoint, and it permits a realist account of things, or at least of the point events out of which physical objects are elaborated. Such an interpretation would make it possible to extend the notion of dependent origination and emptiness to quantum objects—objects that physicists consider to have real existence. Since properties such as spin, position, momentum, and so on, attributed to a micro-object arise in dependence upon its environmental context, we can argue that it is empty by virtue of having no being-in-itself—its being arises in dependence on other beings.

Moreover, the distinction made by Nagarjuna between the indeterminate ground and determinate empirical objects can also be extended to quantum phenomena. If we take the wave function as the analogue of the gestalt function then parallels between the two can be identified. The wave function ascribes no determinate property to a micro-object prior to observation; it is the act of observation that collapses the wave function so that it now ascribes a well-defined property to the micro-object.[31] Thus the wave function prior to observation represents the indeterminate ground state of the micro-entity which constrains the possible results of observation; only after its collapse does the wave function become determinate and give the micro-entity a precisely defined property.

However, Nagarjuna treats the indeterminate ground as somehow associated with a higher reality or a higher truth. By contrast physicists do not consider the wave function to possess a higher reality than the specific properties atomic objects acquire after its collapse. These different orientations of Indian and scientific thinkers can be explained. For Nagarjuna the context of observation is shaped by the theoretical standpoint used to read the configuration—the observed empirical object is a theory-mediated construction elaborated upon the configuring ground. Therefore the empirical object's properties are relative to the theory deployed to interpret the configuration, but the configuration itself is independent of theoretical beliefs. Hence the empirical object is seen as a lower reality than the configuration; and claims about the object are viewed as expressing lower truths than claims about the configuration.

In the case of quantum physics, the wave function describes the state of a real entity in terms of the properties that it can acquire in different observational contexts; after the observation the entity is considered to have acquired a real property—not a property projected onto it by a theoretical orientation. Hence, for quantum physicists both the indeterminate state described by the wave function, and the determinate state after observation, are equally real and express the true state of the object prior to and after observation. They do not express higher and lower truths connected with two levels of reality.

The above discussion also makes evident how paradoxical claims arise when properties that grow in different contexts we have selected as actors are treated as acontextual properties the object had before it enters these contexts, and which then reveal themselves to us simply as spectators. We would then overlook the complementarity of the spectator and actor perspectives Bohr saw as capable of illuminating psychology, and which he treated as a wider extension of quantum complementarity. By similarly ignoring the fact that the position property an electron grows in a measurement context we have set up as actors, we might say that it has a position that we simply discover as spectators. Thus, ignoring our active participation in deciding the context for originating the properties we observe, and treating them as acontextual properties revealed through the contexts we have set up, would tempt us to combine in one picture properties that arise in different mutually exclusive contexts. This shows how taking into account spectator–actor complementarity can illuminate how the insights of complementarity can bring together epistemological perspectives from Buddhist philosophy and quantum theory.[32]

3.4 Complementarity and Dual Function of Theoretical Knowledge

The parallels between Madhyamika and complementarity epistemologies—especially their emphasis on our role as spectators and actors in making scientific observations—suggest that opening a dialogue with Indian philosophy can provide illuminating new directions for contemporary philosophy of science.[33] The significance of such a dialogue can be appreciated even more when we realize that the epistemologies associated with modern science generally see a scientific theory as only concerned with explaining and predicting natural phenomena.[34] However, within the Indian philosophical tradition there has been long recognition of theories as instruments of perception and observation.[35] We have seen that this is the outcome of the importance attached to meditation-induced deautomatization techniques in Indian thought.

However, what can be labeled as postmodern epistemologies associated with thinkers such as Kuhn and Feyerabend, do recognize that all perception is shaped by conception—perception is theory-laden or theory-impregnated so that armed with such knowledge we can read more from sensory stimuli than we otherwise would be able to. For example, armed with theoretical knowledge a geologist can see more in sedimentary layers, a radiologist more in x-ray pictures, and a physicist more in cloud-chamber photographs, than a theoretically uninformed layman. In short, theories enable us to amplify the power of our natural senses by serving as conceptual instruments of perception, in a fashion analogous to the telescope when it is used as a physical instrument of perception.

However, postmodern epistemologies which take into account our active role in observation assume that it requires us to conclude that logic and evidence cannot be sufficient to compare radically different conceptual frameworks. To appreciate this point let us consider the views of one influential postmodern thinker—Richard Rorty.[36] Rorty maintains that with the recognition of the role of theory in perception we have to relinquish the epistemological project of objectively evaluating radically divergent theoretical discourses in science (Rorty 1979: 322–333). Each such discourse can always appeal to its own discourse-constructed evidence to vindicate why it should be preferred to its competitors. Even if we base our evaluations on standard criteria used to compare scientific theories such as inductive strength, confirmatory evidence, experimental corroboration, explanatory scope, and predictive power, we cannot objectively arbitrate across radically divergent scientific discourses. This is because the evidence upon which these criteria are applied is itself read through the theoretical framework being tested. Using Indian philosophical terminology, the evidence invoked involves appeals to thought-forms shaped by the theory under test. This leads Rorty to conclude that the epistemological project of objectively comparing radically different conceptual discourses must collapse.[37]

However, Rorty's claims are questionable. The collapse of epistemology—in the standard sense of the search for foundational representations that can serve as the touchstone for testing all other representations—does not imply that we cannot have objective grounds based upon appeal to logic and evidence for relinquishing one theory in favor of another. Rorty fails to see the possibility that, since theories function as both instruments of inference and instruments of perception, their role as instruments of inference may lead to conclusions that can come into conflict with observations arising from their role as instruments of perception. This would occur when a conflict arises between a prediction made by a theory, deployed as an instrument of inference, and an observation made through the theory, deployed as an instrument of perception. Although such a failure can be deflected by appeal to other auxiliary assumptions to save the theory, these new assumptions also become a part of the theoretical discourse available for use as inferential and observational instruments. They could, in turn, lead to new predictions that clash with observations. Hence the possibility arises of a systematic failure of a theory to function effectively by serving both as an instrument of inference and an instrument of perception in a coherent fashion. Its incapacity to serve these dual functions would provide good grounds for abandoning the theory—especially if a better alternative without similar deficiencies becomes available.

Thus there could be objective reasons for giving up one scientific theory and converting to another. Every theory is subject to constraints beyond its control because it is both an inference tool to provide explanations or make predictions, and a framework through which sensory stimuli are read. Hence, contrary to what Rorty and many other postmodern thinkers assume, logic and evidence can help us to compare theories even if theories function not only as instruments of inference but also as instruments of observation.[38]

However, the role of theories as instruments of perception also raises a second problem that has hardly been addressed systematically by modern philosophers of science—how do we remove the automatic and entrenched tendency to continue to read sensory stimuli in terms of outdated observational theories into which we have been socialized and conditioned over a long period of time? How do we revisualize perceptual experience so that our reading of sensory stimuli becomes informed by the more recent theories we have come to consider as preferable? Such problems arise because learning to read sensory stimuli in terms of a new theory goes beyond conceptual understanding of the theory. This may not be obvious in the case of gestalt shifts of perception—thinking something is "a rabbit" may cause us to see it as a rabbit; and thinking it "a duck" may lead to perception of a duck. But the change may be more difficult to make in many other situations.[39]

Consider the case of a person with good theoretical knowledge of elementary particles and their behavior. Such a person may not be able to read and identify, in cloud chamber photographs tracks of particles—particles such as *alpha* or *beta* particles. The rules for making such identifications have to be learnt over time, but they do not follow directly from theoretical knowledge (even if they are formulated in the context of such knowledge). Moreover the application of these rules, once they have been formulated, may not even require theoretical knowledge. In many centers of high energy physics research, technicians were taught to identify and measure the properties of different particles from the nature and pattern of tracks observed on photographic plates without learning the complex system of physical theories that ultimately makes this possible. Hence, perceiving the world through a theory is a process quite different from learning the theory. It is learning to read sensory stimuli, and make reports, in a language that does not violate the expectations of a theory, but it does not require knowledge of the theory.[40]

The way we embody theories in perception may be understood by way of an analogy—how we build theoretical assumptions into measuring instruments. Scientific instruments have to be standardized prior to their use so that they ignore irrelevant information and read stimuli in an acceptable fashion. What comes out as a direct reading of the instrument is the result of a set of rules, or interpretation functions, built into the instrument. The interpretation functions are not the full-fledged scientific theory, but are formulated in the context of the theory and guided by it.

For example, a pilot's reading of the position of a remote air control tower on the screen of a radar panel is mediated by interpretation functions built into the instrument. Though the reading is a response to external stimuli emanating from the tower, it goes beyond these stimuli—they have been interpreted before being visually presented to the pilot in the form of a display. It is not the case that all electromagnetic and geometric theories have been coded into the instrument—rather the instrument has been set to read stimuli in conformity to some of the assumptions, or implications, of these theories. Such assumptions, or implications, could include information like radio waves travel in straight lines, that they are propagated at a certain velocity, and so on.

In the same way perception may 'embody or model interpretation functions' guided by a theoretical perspective. (Churchland 1979: 39) Indeed most ordinary language is taught in this fashion at the beginning. We are taught to read sensory information in terms of the theoretical suppositions of our culture. We thereby acquire a tacit structuring of the booming buzzing confusion that first confronts us. It is the language of perception into which we are socialized that gives form to these shifting sensations—it enables us to read sensory stimuli to identify objects and make sense of the world.

For most of us the theories that inform our responses to sensory stimuli are often outdated commonsense theories.[41] However, if we could make the most powerful scientific theories guide our readings of sensory stimuli, it would expand considerably our perceptual awareness of the world. One philosopher who has taken seriously the role of deploying our latest theories to expand perceptual consciousness is Paul Churchland. He writes:

> Our current modes of conceptual exploitation (of the natural information contained in our sensations or sensory states) are rooted, in substantial measure, not in the nature of our perceptual environment, nor in the innate features of our psychology, but rather in the structure and context of our common language, and in the process by which each child acquires the normal use of that language ... But our current conceptual framework is just the latest stage in the long evolution process that produced it, and we may examine with profit the possibility that perception might take place within the matrix of a different and more powerful conceptual framework. The obvious candidate here is the

> conceptual framework of modern physical theory—of physics, chemistry, and their many satellite sciences. That the conceptual framework of these sciences is immensely powerful is beyond argument, and its credentials as a systematic representation of reality are unparalleled. It must be a dull man indeed whose appetite will not be whet by the possibility of perceiving the world directly in its terms. (Churchland 1979: 7)[42]

However, Churchland fails to confront one serious obstacle that lies in the path of such a transformation of our perceptual consciousness. Reconditioning the ways in which we read sensory stimuli is not as easy as merely learning a new theory. Perceptual sets are more resistant to change than theoretical beliefs, which can be easily shifted. Just as an instrument would give the same readings even if we happen to discard the theories that guide its interpretation function, so would our perceptual sets continue to be shaped by interpretation functions built into them by earlier discarded beliefs that conditioned them—unless we can actively intervene to alter these functions by deconditioning our perceptual sets. Yet Churchland fails to tell us how we can effectively rid ourselves of perceptual sets into which we have become conditioned, and replace them with new ones more in accord with our latest theoretical beliefs. Though he recognizes the problem of changing perceptual sets he attributes it to cultural inertia, which he thinks can be overcome only gradually.

I would like to suggest that meditation techniques can give technological teeth to bringing about changes in perceptual sets envisaged by Churchland. They offer us psychosomatic techniques to decondition perceptual sets shaped by outdated theories and can recondition us into new better perceptual sets shaped by currently preferred theories. Indeed this is precisely what, as we have seen, they are designed to do. The deconditioning process involves deautomatization techniques that, as we saw earlier, render perception plastic by suspending conceptually constructed perceptual sets. The reconditioning technique involves visualization strategies that train us to read sensory stimuli through other new theories and entrench them as ways of experiencing the world.[43]

Indeed it could be argued that when we make a conceptual change we only make half a cognitive change; the change is not complete until it leads to a difference in our experience of the world. A philosophy that confines itself to only the conceptual level is noetically limited in scope. In this regard, the orientation of modern philosophy concerned with only conceptual analysis and criticism can be enriched by insights from Indian philosophy. Most Indian philosophers have traditionally seen conceptual change to be but a stage in a more radical transformation of cognition. This is reflected in the term "*darsana*" (meaning "to envision" in Sanskrit) used to refer to Indian philosophical traditions. *Darsana* is often (mis) taken to be the Indian equivalent of the modern sense of a worldview.[44] However Indian schools have generally recognized that "envisioning" involves a cognitive change that goes beyond mere conceptual change in understanding—it also demands an experiential transformation. By treating their various schools of philosophy as *darsanas*, Indian philosophers see a philosophical position as becoming completely understood only after the change of belief is completed by a belief-guided revisioning of the learner's perceptual sensibilities. Reconceptualization and revisualization are seen as two stages of any significant cognitive change.

It is precisely for this reason that understanding and deploying meditation techniques can make a significant contribution to enriching modern philosophy of science. Meditation practices not only facilitate deautomatizing outdated conceptual categories conditioning our perceptual experience of the world, but also give revisualization techniques that take new beliefs further by entrenching them into our perceptual sensibilities. Indeed Indian philosophical traditions in Buddhist, Hindu, and Jain schools have always stressed such envisioning of correct beliefs (*darsana*) as integral to their philosophical training. Hence, meditation has always traditionally been an integral component of their philosophical training and educational practices.

One major obstacle to embracing this wider conception of philosophy that is able to integrate reconceptualization and revisualization is the prejudice that philosophy at its best is a theoretical enterprise, and that to treat it as also involving the practical task of transforming vision is somehow contaminating. Indeed this has even led many modern writers to deny the existence of Indian traditions of philosophy altogether. They argue that the practical and soteriological task Indian philosophers set themselves, over and above pure theoretical pursuits, somehow diminishes their approach to knowledge. These writers even suggest that the Indian approaches are not strictly philosophical. It is such a narrow vision of philosophy as a purely theoretical enterprise that inspires the phenomenologist Edmund Husserl. In his highly influential Vienna lecture of 1935, entitled "Philosophy and the Crisis of Humanity", Husserl repudiates the notion of the existence of Eastern philosophy precisely on these grounds:

> Today we have a plethora of works about Indian philosophy, Chinese philosophy, etc., in which these are placed on a plane with Greek philosophy and are taken as merely different historical forms under one and the same idea of culture. Naturally, common features are not lacking. ... In both cases one may notice a world-encompassing interest that leads on both sides—thus also in Indian, Chinese, and similar 'philosophies'—to universal knowledge of the world. ... But only in the Greeks do we have a universal ('cosmological') life-interest in the essentially new form of a purely 'theoretical' attitude ... [to] bring about *theoria* and nothing but *theoria*. (Husserl 1970: 280)[45]

The same argument also leads Husserl to reject the notion that there has been any Eastern science:

> [I]t is a mistake, a falsification of their sense, for those raised in the scientific ways of thinking created in Greece and developed in the modern period to speak of Indian and Chinese philosophy and science. (Husserl 1970: 284–285)

Husserl's conception of philosophy as theoretical knowledge, and such knowledge alone, is narrow, questionable, and damaging to the enterprise of philosophy itself. By confining itself to purely theoretical knowledge, philosophy cannot even adequately address the epistemological issues it confronts as a result of the theory impregnation of experience.

Husserl's confining vision of philosophy would also limit our conception of science, since the theory-laden nature of perceptual experience raises epistemological issues that cannot be adequately resolved without taking into account the techniques involved in deautomatizing and revisualizing our cognitive sensibilities. His vision of philosophy and science as *theoria* would emasculate us within a framework for philosophy that fails to see the role of theories as instruments of perception. Such a vision was not even a part of Greek science and philosophy as Husserl presumes, since Plato, and the Neoplatonists who followed him, were far more sensitive to the need for personal transformation as a complement to theoretical knowledge.[46] Neither is it sensitive to Bohr's view that in psychology we are both spectators and actors in the drama of existence.

However, the role of knowledge as *theoria* alone has another limitation. It leads to the notion that only knowledge carried by humans is important, and that knowledge embodied in natural systems is valuable only to the extent that it becomes known to humans. It ignores the fact that human knowledge is only one part of a broader system of knowledge, which includes information carried by ecosystems in the biosphere, that shape processes in nature. This limitation of modern epistemology becomes evident when we look at the parallels between complementarity and Daoist epistemologies noticed by Bohr. We will find in the following chapter that Daoist epistemology offers another way of enriching current philosophy of science by going beyond the conception of knowledge as simply *theoria*, and that this also has implications for formulating economic theories that can take into account the self-regulating processes of natural systems in the biosphere.

Notes

1. The significance of such an approach has been stressed by Sarukkai (2005). He argues that Indian approaches to mathematics, logic, and ontology can cast new perspectives on current philosophies of science.
2. Gestalt theories of perception have been criticized for being descriptive rather than explanatory by many cognitive psychologists and neuroscientists who see the approach as redundant and uninformative. Thus Bruce, Green, and Georgeson write:

   > The physiological theory of the gestaltists has fallen by the wayside, leaving us with a set of descriptive principles, but without a model of perceptual processing. Indeed, some of their "laws" of perceptual organization today sound vague and inadequate. What is meant by a "good" or "simple" shape, for example? (Bruce et al. 1996: 110)

   However, such a conclusion seems unusually restrictive in its cognitivist and information processing view of perception, and has been criticized by postcognitivists who emphasize a non-reductionist approach to visual perception. See Gibson (1979) and Gibson and Pick (2000).
3. For a recent review of the role of theory-ladenness in the philosophy of science, see Schindler (2013).
4. See Olson (2002).
5. There is some experimental evidence for this based on studies of the effects of meditation. These show that meditation practice involves reducing the conceptual interpretation of sensory stimuli so that they come to be perceived more directly without the mediation of interpretive influences. Studying the influence of mindfulness meditation on visual sensitivity to lowered threshold of light stimulation Brown et al. (1984) conclude that quieting some of the higher mental processes that normally obstruct the perception of subtle events led to better rate of detection of single light

flashes. Similarly the study of the perception of visual illusions by novice and long-term meditators lead Tloczynski et al. (2000) to conclude:

> A person who meditates consequently perceives objects more as directly experienced stimuli and less as concepts ... With the removal or minimization of cognitive stimuli and generally increasing awareness, meditation can therefore influence both the quality (accuracy) and quantity (detection) of perception.

6. Speaking of the Madhyamika Buddhist experience of the absolute Murti writes that the absolute is "beyond the scope of discursive thought, language and empirical activity" and is, therefore, unthinkable, unutterable, and unteachable. It can only be directly experienced (Murti 1960: 244).
7. Attempts to develop a constructivist account of the mystical (intuitive) experience lend further support to this approach though in a somewhat indirect fashion. According to Katz the phenomenology of the mystic (intuitive) experience is radically shaped by prior conceptual beliefs brought to bear on the experience (Katz 1978: 22–74). Gimello also argues that the experience is the result of a psychosomatic enhancement of our beliefs. (Gimello 1983: 85) Such constructivist accounts lend credence to the view that the intuitive experience, albeit very different in many ways from ordinary experience, is nevertheless an emanation of the same powers and faculties that are deployed in the construction of everyday experience. See also Hollenback (2007) which explores how the mystic experience in different religious traditions is shaped by historical and cultural contexts that influence both the perceptual and affective content of the experience.
8. For a more detailed discussion of this problem and related issues see Chap. 3 "The Cognitive Status of the Mystical Experience" in Wainwright (1981: 82–137). See also Huxley (1946: 5). In recent years there has been increasing interest in the evolutionary basis of religious thinking. See Barrett (2004), Boyer (2001), and Tremlin (2010).
9. Deikman (1963) in Tart (1972: 205). Two studies by Deikman (1963 and 1966) on experimental meditation are reprinted in Tart (1972).
10. Deikman (1963) quoted in Tart (1972: 212).
11. Deikman (1966) in Tart (1972: 32–37).
12. Loy (1988: 87). Loy adds that in the Western tradition, despite Hume, it is often only the ontological status of the object, but not the subject, which is questioned.

13. Dignaga was one of the founders of Indian logic. *Hetucakra* or *Wheel of Reason* introduced a new form of deductive reasoning and constituted his first work on formal logic. His other works include *The Treatise on the Objects of Cognition (Ālambanaparīk ṣā), The Treatise on Systems of Cognition (Pramāṇa-samuccaya)*, and *The Treatise on the Correct Principles of Logic (Nyāya-mukha)*.

    Dharmakirti was a follower of Dignaga who built upon his work, as well as pioneered new directions in Indian and Buddhist logic. He was also a teacher at the renowned Nalanda University whose works are still studied as a part of the monastic curriculum in Tibet. See Dreyfus (1997).

    It is important to note that the logical studies of these scholars both built on a long tradition of Indian logic from the time of Medhatithi Gautama (c. sixth century BCE), Pāṇini (c. fifth century BCE), the Vaisheshika school's analysis of atomism (c. second century BCE), the analysis of inference by Gotama (c. second century) who was the founder of the Nyaya school of Hindu philosophy, and the tetralemma of Nagarjuna (c. second century CE). Their work in turn influenced the Navya-Nyaya school of logic through to early modern times.
14. However, two significant differences preclude the complete identification of the views of Indian thinkers and the gestalt psychologists. First, the early gestalt psychologists adopted a physicalist rather than a sensationalist account of the basic stimuli from which percepts were constructed—these were retinal stimulations rather than sensations. Second, they saw the organizing principles that generated percepts as built into the human nervous system; or as isomorphic to electrical fields in the brain that developed in accordance with inherited gestalt laws. Though they did allow scope for past experience to influence perception, this was conceded only to a very limited extent. (Gregory 1970: 10) By contrast, in Indian tradition the conceptual frameworks brought to bear on independently given sensations radically condition perceptual organization. For a more thorough review of the role of gestalt psychology in the genesis of modern theories of perception refer to Uttal (1988), especially the chapter on "Theories of Form Perception", pp. 49–107.
15. Mansfield (1989) argues that we need to open a dialogue between quantum theory and Madhyamika Buddhism because of their close philosophical affinity. Varela et al. (1991) examined the epistemic parallels between Buddhism, quantum theory, cognitive psychology, and biology and their connections with postmodern philosophies (especially in relation to the views of Rorty). More recently Zajonc (2004) brings together leading

physicists, a historian, and a Buddhist thinker to explore connections between quantum physics and Buddhist philosophy. These studies continue Bohr's recognition of the epistemological parallels between Buddhism and quantum physics. An early attempt at explaining these parallels can be found in Balasubramaniam (1992).

16. Nagarjuna (c. 150–250 CE) is closely linked with the Nalanda Buddhist University in India. In its heyday from the fifth to thirteenth centuries it brought together scholars from India and China, as well as Central Asia and Southeast Asia. The significant influence of Buddhism in India and China provoked the counter reaction from Hindu thinkers in India and Confucian scholars in China reflected in the Vedanta philosophy of Shankara (early eighth century) and neo-Confucian response of Zhu Xi (1130–1200), both of which absorbed the doctrines of Madhyamaka while also attempting to counter it.
17. Indeed this is a general, albeit not universal, tendency in all Indian philosophy. According to Matilal "whatever might have been the motive or driving force behind this refutation of [the] external material world, it was received with all philosophic seriousness in India" (Matilal 1974: 155).
18. That Nagarjuna's view is that the world of objects is constructed by the intellect is stressed by Ramanan:

> The world of convention is called *nirmana* to indicate that it is a creation; it is called *samvrti* to indicate that it veils the truth of things; it is called *vyarahara* to say that is has mundane truth, "empirical validity", although devoid of ultimacy; it is called *prapanca* to show that it is an elaboration through concepts and conventional entities. The "builder" of the world is *vijnana* or *citta* as a self-conscious principle of intellection. And in this building of the world the two, *nama* and *laksana*, names and what they stand for, constitute the warp and the woof. (Ramanan 1978: 73)

David Loy argues that the Tibetan and Chinese Madhyamika exegetical traditions perceive the relation between *vikalpa* and *prapanca* as that between the subjective mental act of conceptualization and its crystallized counterpart experienced as objective. He argues that *prapanca* might be defined as the differentiation of the nondual world of *nirvikalpa* experience as discrete objects of the phenomenal-world by virtue of *savikalpa* thought-construction (Loy 1988: 53–54).

19. There have been many translations of the *Mulamadhyamakakarika*. They include Garfield (1995), Kalupahana (1986), Sprung (1979), and Inada (1970).

20. This doctrine of two truths has parallels in Greek Pyrrhonism as noted by McEvilley (2001: 474) and Conze (1959: 244).
21. Gunaratne (1986: 213) characterizes it as "one of the most perplexing problems in the study of Buddhist thought." See also Priest (2010), Stcherbatsky (2008), Wayman (1997) and Westerhoff (2006).
22. As with the two truths doctrine there are also parallels in Greek thought to the logic of the *catuskoti*. This has been noted by McEvilley:

    An extraordinary similarity, that has long been noticed, between Pyrrhonism and Mādhyamika is the formula known in connection with Buddhism as the fourfold negation (catuṣkoṭi) and which in Pyrrhonic form might be called the fourfold indeterminacy. (McEvilley 2001: 495)

23. Robinson (1967: 50) argues that Nagarjuna did not reject any laws of thought.
24. The *anekāntavāda* is central to the philosophy of Jainism, which maintains that truth and reality are always perceived from diverse points of view and that no single point of view can be taken to give the complete truth. It is often taught through the parable of the blind men and the elephant, where each blind man feeling a different part of an elephant (trunk, ear, leg, etc.) mistakenly assumes that its shape is the shape of the whole elephant. This is meant to illustrate the Jain claim that no single human viewpoint can represent absolute truth. Closely linked to the non-onesidedness doctrine is the Jain theory of sevenfold prediction (*syādvāda*).
25. For a more detailed study of Madhyamika dialectics, see Ghose (1987) and for a more recent study, see Tillemans (2013).
26. According to Murti, "the Absolute in itself is indeterminate (*sunya*); no category of thought applies to it. It is ignorance (*avidya*) that invests it with the colorful forms that we come across in ordinary experience." (Murti 1960: 238).
27. Modern nonfoundationalists like Rorty and Putnam are also sensitive to the way we construct worlds from sensory stimuli. For a more specific study of these intimate links between the Madhyamika, modern cognitive science, and nonfoundationalism see Varela et.al. (1991), especially Chap. 10 on "The Middle Way" which contrasts the nonfoundational views of Heidegger, Rorty, Goodman and Putnam with the Buddhist notion of groundlessness.
28. This builds on the distinction between epistemic duomorphism and monomorphism made by Forman (1989). See also Forman (1999).
29. Murti emphasizes that the general duomorphic structure of the Madhyamika is applicable for all alternatives and not just for our claims with regard to empirical objects:

> Four alternatives are possible on any subject. The basic alternatives are two: Being and Non-Being, Affirmation and Negation. From these, two others are derived by affirming or denying both at once: both Being and Nonbeing, and neither Being nor Non-Being. It may be thought that in avoiding the two extremes, the Madhyamika takes a middle position in between the two. No; he does not hold any middle position. Or, the middle is no position; it is beyond concept or speech; it is the transcendental, being a review of all things. (Murti 1960: 129)

30. The analogy between visual perception and quantum measurement proposed seems to be somewhat supported by G. Szamosi in a paper entitled "Naturalizing the Copenhagen Interpretation". In the abstract, Szamosi summarizes the paper as follows:

> Consider two simple textbook observations: a) quantum measurement is an information processing method invented for the purpose of exploring domains of the external world which are not accessible otherwise; b) quantum measurement interprets signals from the external world with the help of a computing algorithm invented specifically for this purpose. Replace the words "quantum measurement" and "invented" in a) and b) by "vision" and "evolved" respectively. Both sentences remain simple textbook observations; this time about vision and its biological evolution. Since both a) and b) are substantial statements whether they refer to vision or to quantum measurement the analogy between these processes suggests that quantum measurement may be viewed as a culturally evolved perceptual mode. (Szamosi 1993: 305)

This leads Szamosi to conclude that we can have a realistic formulation of the Copenhagen interpretation which also explains why, in quantum measurement, some observables exist in nature only when observed.

Similarly, Anton Amann (1993) notes that quantum mechanics and gestalt psychology exhibit isomorphic conceptions and problems—the property of a quantum object depends on its interaction with the environment such as a measurement apparatus analogous to the property of a gestalt object; also quantum phenomena and gestalt perception are organized in holistic ways.

31. However the analogy can only be developed to a limited extent. The gestalt function exists as a real configuration; the ontological status of the wave function is open to debate. See Jammer (1974), especially Chap. 2.
32. We could also do the same for the Jain *syavada* and extend it to illuminate the quantum situation. Then we could say that the micro-entity before measurement is a particle, wave, indeterminate, both particle and wave, particle and indeterminate, wave and indeterminate, wave-particle and indeterminate. Each of these are judgments made from a single or a multiplicity of contexts.

33. See Mansfield (1989), Balasubramaniam (1992), and Zajonc (2004).
34. This implies that a theory is seen as an instrument of inference that allows us to logically deduce the results of observation—albeit in conjunction with other auxiliary theories, and the specification of the initial conditions that describe the circumstances in which the theory is being applied. For inductivists, positivists, logical empiricists, and falsificationists, testing a theory essentially involves checking inferences from the theory against the results of experiment or observation and, depending on their epistemological and methodological orientations, deciding whether observations implied, verified, increased the degree of confirmation, or falsified the theory. Testing a theory involved looking at the relationship between the theory and data—not looking at the theory in relation to other theories, which were its competitors.

    The exception here is Kant. But even Kant assumed his categories to be foundational though he recognized their role in shaping perceptual experience. It is really with the neo-Kantians—beginning with Hegel—that we see the turn toward non-foundational views of perception which today have grown into the full-fledged set of doctrines characterized as postmodern.
35. The notion of theories as instruments of observation in the sciences is associated with the views of Hanson (1958), Toulmin (1990), Kuhn (1970), and Feyerabend (1978). The role of theories as observational instruments is precisely what makes us actors, and not simply spectators, in making scientific observations. This raises epistemological issues that have yet to be resolved in current philosophy of science. The Enlightenment epistemological tradition did not face this problem because it treats the evidence of perception as foundational for theoretical knowledge, since it takes it to be purely something imposed upon us as spectators and to be uniformed by theoretical conceptions.
36. Rorty links his antifoundational and antiepistemology views with the deconstructive work of Dewey, Heidegger and Wittgenstein. See Rorty (1979), especially pp. 5–7.
37. It also leads Rorty to question the viability of the enterprise of comparative philosophy. Rorty argues that such comparisons cannot be fruitful because the very notion of philosophy itself is different across cultures so that no dialogical exchange can influence or profit those coming from radically different traditions. Thus Rorty's postmodern claim puts into question Bohr's attempt to make complementarity epistemology a new framework for science that would also open the door to intercultural epistemological dialogue questionable on two counts. First, it rejects Bohr's notion of an objective epistemology for science. Second, it rejects the possibility and utility of engaging Eastern philosophies as a way of advancing epistemological

understanding of current science. For a more elaborate discussion of Rorty on comparative issues, see Zhang (2007).

38. Both New Age and Indian thinkers who argue that intuitive experience reveals the objects in the world as simply psychological constructions, or thought-forms, also fail to see the dual function of theories. Otherwise they would be led to conclude that some of our mental constructions may turn out to be illusory when we discard a theory as an instrument of observation because it does not also effectively function as an instrument of inference and vice versa—a conclusion that would put into question the psychological idealism that our theories somehow create *de novo* the objects in the world.

39. Take for example experiments with inverting lenses invoked by Thomas Kuhn (1970: 112) to illustrate this point most dramatically. A subject is made to wear goggles with inverting lenses that make the world appear upside down, although the subject knows intellectually that this is not the case. The subject is initially disoriented but after a while begins to adjust as the visual world becomes less confused, and finally flips over so that it appears just as it did prior to putting on the goggles. What is important is that the change in perception is not instantaneous but takes place over time. Mere conceptual knowledge that the world is not upside down does not immediately yield the correct visual image when the subject puts on the inverting goggles—only after prolonged use of the goggles does the visual image begin to converge and conform to the subject's conceptual expectations. Moreover, when the goggles are removed the world again appears upside-down even though the subject knows this to be an illusion. It takes a further interval of time for the visual field to renormalize. These experiments reveal that a period of time is needed by the subject to learn to read sensory stimuli guided by correct conceptual beliefs so that perceptual experience comes to match conceptual beliefs.

    The earliest inverting lenses experiments were conducted by George Stratton in the late nineteenth century, see Stratton (1897).

40. In early particle detection experiments, from the 1920s to the 1950s, the instruments deployed were cloud chambers to be replaced later by bubble chambers. Cloud chambers contained a supersaturated vapor of water or alcohol, which condensed when it was ionized by a charged particle, such as an alpha or beta particle, passing through it. The trail of the particle can be seen along the ionization tracks that have distinctive shapes depending on the charge and mass of the particle. These tracks can also be made to curve by the application of a magnetic field across the chamber so that their curvature and direction of deflection would also give further information about the nature, mass, and charge of particles. To a large extent these fundamental detection methods pioneered by cloud chambers continue to play a role in later particle detectors of different designs. For a history of particle detection see Grupen and Buvat (2012).

41. We continue to say "the sun rises" and "heat flows" even though the theories that motivated these assertions are now defunct.
42. In a more recent study Churchland (2012) attempts to give a neurobiological grounding for this thesis. He shows how such an account provides a systematic understanding of the way low level epistemological activities in the brain are integrated into the wider framework of language structures and regulatory mechanisms at the social level. This biological grounding serves to further consolidate his thesis that conceptual frameworks can serve to amplify perceptual processes.
43. In the philosophy of science, such revisualization of experience that accompanies reconceptualization of thought has been emphasized by Kuhn when he speaks of paradigm changes as gestalt switches. What Kuhn sees as gestalts have been labeled in Tibetan Buddhism as *tulpa*. The term *tulpa* was rendered into English as 'thoughtform' by Evans-Wentz (1954: 29) who writes:

    > In as much as the mind creates the world of appearances, it can create any particular object desired. The process consists of giving palpable being to a visualization, in very much the same manner as an architect gives concrete expression in three dimensions to his abstract concepts after first having given them expression in the two-dimensions of his blue-print.

44. The term '*darsana*' in Indian philosophy refers to the way each philosophical system is designed to shape not only our conceptual understanding but also our perceptual and affective experience of the world. It is traditionally assumed that there are different schools of darsanas such as *Samkhya, Yoga, Nyaya, Vaishesika, Mimamsa, and Vedanta* in the Hindu tradition, as well as Buddhist, Jain, and the materialist Carvaka schools.
45. See Mohanty (1993: 282–292) for a critique of this Husserlian Eurocentrism.
46. The similarities between the Greek idea of *theoria* and the Indian idea of *darsana* has been noted by Rutherford (2000).

Bibliography

Amann, Anton. 1993. The Gestalt Problem in Quantum Theory: Generation of Molecular Shape by the Environment. *Synthese* 97(1): 125–156.

Balasubramaniam, Arun. 1992. Explaining Strange Parallels: The Case of Quantum Mechanics and Mādhyamika Buddhism. *International Philosophical Quarterly* 32(2): 205–223.

Barrett, Justin L. 2004. *Why Would Anyone Believe in God?* Chicago: AltaMira Press.

Bhatt, S.R. 2000. Logic and Language in Buddhism. In *Companion Encyclopedia of Asian Philosophy*, eds. Carr and Mahalingam, 414–434. Abingdon: Routledge.

Bohr, Niels. 1958. *Atomic Physics and Human Knowledge*. New York: John Wiley & Sons.

Boyer, Pascal. 2001. *Religion Explained: The Evolutionary Origins of Religious Thought*. New York: Basic Books.

Brown, Daniel, Michael Forte, and Michael Dysart. 1984. Differences in Visual Sensitivity Among Mindfulness Meditators and Non-meditators. *Perceptual & Motor Skills* 58: 727–733.

Bruce, V., P. Green, and M. Georgeson. 1996. *Visual Perception: Physiology, Psychology and Ecology*. Hove: LEA.

Carr, Brian, and Indira Mahalingam, eds. 2000. *Companion Encyclopedia of Asian Philosophy*. London: Routledge.

Churchland, Paul M. 1979. *Scientific Realism and the Plasticity of Mind*. Cambridge, MA: Cambridge University Press.

Churchland, Paul M 2012. *Plato's Camera: How the Physical Brain Captures a Landscape of Abstract Universals*. Cambridge, MA: MIT Press.

Conze, Edwards. 1959. *Buddhism: Its Essence and Development*. New York, USA: Harper and Row.

Deikman, Arthur. 1963. Experimental Meditation. *Journal of Nervous and Mental Diseases* 136: 329–343.

———. 1966. Deautomatization and the Mystic Experience. *Psychiatry* 29: 324–338.

Dreyfus, Georges. 1997. *Recognizing Reality: Dharmakirti's Philosophy and Its Tibetan Interpretations*. New York: State University of New York Press.

Evans-Wentz, W.Y., ed. 1954. *The Tibetan Book of the Great Liberation*. Oxford: Oxford University Press.

Forman, Robert K. 1989. Paramartha and Modern Constructivists on Mysticism: Epistemological Monomorphism Versus Duomorphism. *Philosophy East and West* 39: 393–418.

Forman, Robert K. 1999. *Mysticism, Mind, Consciousness*. New York: State University of New York Press.

Freud, Sigmund. 1949. *Psychopathology of Everyday Life*. Trans. A.A. Brill. London: Ernest Benh Ltd.

Garfield, Jay L. 1995. *The Fundamental Wisdom of the Middle Way*. Oxford: Oxford University Press.

Ghose, Ramendranath. 1987. *The Dialectics of Nagarjuna*. Allahabad, India: Vohra Publishers and Distributors.

Gibson, James J. 1979. *The Ecological Approach to Visual Perception*. Boston: Houghton Mifflin.

Gibson, Eleanor, and Anne Pick. 2000. *An Ecological Approach to Perceptual Learning and Development*. Oxford: Oxford University Press.

Gimello, Robert M. 1983. Mysticism in Its Contexts. In *Mysticism and Religious Traditions*, ed. Steven T. Katz, 1983: 61–88. Oxford: Oxford University Press.

Gregory, Richard L. 1970. *The Intelligent Eye*. London: Weidenfeld & Nicolson.

Grupen, Claus, and Irene Buvat, eds. 2012. *Handbook of Particle Detection and Imaging*. Berlin: Springer.

Gunaratne, R.D. 1986. Understanding Nagarjuna's Catuskoti. *Philosophy East and West* 36(3): 213–234.

Happold, F.C. 1984. *Mysticism: A Study and an Anthology*. New York: Penguin Books.

Hollenback, Jess. 2007. *Mysticism: Experience, Response, and Empowerment*. University Park: Penn State University Press.

Husserl, Edmund. 1970. *The Crisis of European Sciences and Transcendental Phenomenology: An Introduction to Phenomenological Philosophy*. Translated with an introduction by David Carr. Evanston: Northwestern University Press.

Huxley, Aldous. 1946. *The Perennial Philosophy*. London: Oxford University Press.

Inada, Kenneth. 1970. *Nagarjuna: A Translation of His Mulamadhyamikakarika*. Tokyo: The Hokuseido Press.

Jammer, Max. 1974. *The Philosophy of Quantum Mechanics: The Interpretation of Quantum Mechanics in Historical Perspective*. New York: John Wiley & Sons.

Kalupahana, David J. 1986. *Nagarjuna: The Philosophy of the Middle Way*. New York: State University of New York Press.

Katz, Steven T., ed. 1978. *Mysticism and Philosophical Analysis*. New York: Oxford University Press.

Katz, Steven T., ed. 1983. *Mysticism and Religious Traditions*. New York: Oxford University Press.

Kuhn, Thomas. 1970. *The Structure of Scientific Revolutions*. Chicago: University of Chicago Press.

Loy, David. 1988. *Nonduality: A Study in Comparative Philosophy*. New Haven & London: Yale University Press.

Mansfield, Victor. 1989. Madhyamika Buddhism and Quantum Mechanics: Beginning a Dialogue. *International Philosophical Quarterly* 29: 371–391.

Matilal, B.K. 1974. A Critique of Buddhist Idealism. In *Buddhist Studies in Honour of I.B. Horner*, eds. L. Cousins, A. Kunst, and K.R. Norman, 139–169. Dordrecht, Holland: D. Reidel.

McEvilley, Thomas. 2001. *The Shape of Ancient Thought: Comparative Studies in Greek and Indian Philosophies*. New York: Allworth Press.

Mohanty, Jitendra Nath. 1993. *Reason and Tradition in Indian Thought: An Essay on the Nature of Indian Philosophical Thinking*. USA: Oxford University Press.

Murti, T.R.V. 1960. *The Central Philosophy of Buddhism: A Study of the Madhyamika System*. London: George Allen and Unwin Ltd.

Olson, Carl. 2002. *Indian Philosophers and Postmodern Thinkers: Dialogues on the Margins of Culture*. Oxford: Oxford University Press.

Priest, Graham. 2010. The Logic of the Catuskoti. *Comparative Philosophy* 1(2): 24–54.

Ram-Prasad, Chakravarthi. 2005. *Eastern Philosophy*. London: Cassell.

Ramanan, Venkata. 1978. *Nagarjuna's Philosophy*. Delhi: Motilal Banarsidass.

Robinson, Richard H. 1967. *Early Madhyamika in India and China*. Madison, WI: The University of Wisconsin Press.

Rorty, Richard. 1979. *Philosophy and the Mirror of Nature*. Princeton, New Jersey: Princeton University Press.

Ruegg, D. Seyfort. 1981. *The Literature of the Madhyamaka School of Philosophy in India*. Wiesbaden: Harrassowitz.

Rutherford, Ian. 2000. *Theoria* and *Darshan*: Pilgrimage as Gaze in Greece and India. *Classical Quarterly* 50: 133–146.

Samten, Geshe Ngawang and Jay L. Garfield. (2006). "Translators' Introduction" in Rje Tsong Khapa's *Ocean of Reasoning: A Great Commentary on Nagarjuna's Mulamadhyamakakarika*. Oxford: Oxford University Press.

Sarukkai, Sundar. 2005. *Indian Philosophy and Philosophy of Science*. New Delhi: Centre for Studies in Civilizations.

Schindler, Samuel. 2013. Theory-laden Experimentation. *Studies in History and Philosophy of Science* 44(1): 89–101.

Schumacher, John, and Robert Anderson. 1979. In Defense of Mystical Science. *Philosophy East and West* 29: 73–90.

Sharma, Chandradhar. 1976. *A Critical Survey of Indian Philosophy*. Delhi: Motilal Banarsidass.

Smart, Ninian. 2000. *World Philosophies*. London & New York: Routledge.

Sprung, Mervyn. 1979. *Lucid Exposition of the Middle Way*. Boulder: Prajna Press.

Stcherbatsky, T.H. 2008. *Buddhist Logic*. (2 volumes). New Delhi: Motilal Banarsidass Publishers.

Stratton, George M. 1897. Upright Vision and the Retinal Image. *Psychological Review* 4: 182–187.

Streng, Frederick. 1967. *Emptiness: A Study in Religious Meaning*. New York: Abingdon Press.

Szamosi, G. 1993. Naturalizing the Copenhagen Interpretation. *Dialectica* 47(4): 305–325.

Tart, Charles T. 1972. *Altered States of Consciousness*. New York: John Wiley & Sons.

Tillemans, Tom. 2013. "How Do the Madhyamikas Think?" Revisited. *Philosophy East & West* 63(3): 417–425.

Tloczynski, Joseph, Aimee Santucci, and Eileen Astor-Stetson. 2000. Perception of Visual Illusions by Novice and Longer-Term Meditators. *Perceptual and Motor Skills* 91(3 Pt 1): 1021–1026.

Tremlin, Todd. 2010. *Minds and Gods: The Cognitive Foundations of Religion*. Oxford: Oxford University Press.

Uttal, William R. 1988. *On Seeing Forms*. London: Lawrence Erlbaum Associates.

Varela, Francisco, Evan Thompson, and Eleanor Rosch. 1991. *The Embodied Mind:Cognitive Science and Human Experience*. Cambridge, MA: The MIT Press.

Wainwright, William J. 1981. *Mysticism: A Study of Its Nature, Cognitive Value and Moral Implications*. Great Britain: The Harvester Press.

Wayman, Alex. 1977. Who Understands the Four Alternatives of the Buddhist Texts? *Philosophy East and West* 27(1): 3–21.

Westerhoff, J. 2006. Nagarjuna's Catuskoti. *Journal of Indian Philosophy* 34(4): 367–395.

Zajonc, Arthur, ed. 2004. *The New Physics and Cosmology: Dialogues with the Dalai Lama*. Oxford: Oxford University Press.

Zhang, Wei. 2007. *Heidegger, Rorty, and the Eastern Thinkers: A Hermeneutics of Cross-Cultural Understanding*. New York: SUNY Press.

CHAPTER 4

Anthropological Complementarity of the Natural and Cultural

4.1 Yin-Yang Complementarity in Chinese Thought

In the last chapter we found that Bohr's extension of complementarity into psychology and his discovery of parallels to complementarity in Buddhist philosophical traditions are interconnected. Complementarity in psychology arises because of the gestalt structure of perception, and Indian thinkers are responding to gestalt experiences. We also discovered that this made it possible to use Indian epistemological views to illuminate current concerns in the philosophy of science. Similarly, we will find that Bohr's attempt to extend complementarity into the social sciences, and his discovery of parallels in Daoist thought, are interconnected. We will find that this not only has implications for supporting recent eco-feminist critiques of Bacon's experimental method for science, but also ecological critiques of standard economic theory. As a consequence we will discover that it is possible to use Chinese Daoist epistemological views to illuminate both ecofeminist and eco-economic concerns in the philosophy of science.

Central to the extension of complementarity into the social sciences proposed by Bohr is its parallel to the complementarity of the feminine and masculine principles, or *yin-yang* complementarity, in Chinese thought.[1] Bohr recognized this relationship, as we have seen, when he chose the Chinese symbol of *taijitu* (Unity of Yin-Yang) as the crest for his shield when he was knighted. Such a relationship will not appear surprising once we recognize that quantum complementarity is a response to grown

Arun Bala, *Complementarity Beyond Physics*,
DOI 10.1007/978-3-319-39784-9_4

properties in the atomic domain, and the dominant natural philosophy of ancient China is what has been characterized by Joseph Needham as 'organismic materialism'—a philosophy of nature modeled upon organic systems in which grown properties are ubiquitous.

In order to appreciate the structure of Chinese organic thought, it is illuminating to compare it with the ancient Greek tradition of natural philosophy—at least in the most developed form it reached with Aristotle. Although Aristotle, like the Chinese, also saw nature as an organic system, the two views are quite different. This is most clearly seen when we compare the Greek and Chinese conceptions of causal relations in the world. Aristotle explains natural phenomena by appealing to essential causes within objects or agents, but the Chinese by invoking correlative causes dependent on the relations of one thing to other things in its environment. This difference between the Greek and Chinese views has profound implications for how the fundamental problems of philosophy are construed in the two traditions, and it shows why complementarity arises in Chinese but not Greek organic thought.

Let us take Aristotle first. Aristotle sees phenomena as conditioned by the action of four types of causes he labels as final, material, formal, and efficient. Furthermore, he grounds these causes themselves on the more fundamental concepts of matter and form that he uses to analyze all processes of change. The action of these causes can be illustrated by looking at the example of how Aristotle explains the growth of an oak tree from an acorn. He analyzes the process as follows. The final cause of the oak is the purpose for which it was grown; the material cause is the soil, water, and other things needed to nurture its growth; the formal cause is the form of the tree that exists potentially in the acorn; and the efficient cause is the person who planted the acorn. In the case of an oak growing in the wild it is a bit more problematic for Aristotle to give final and efficient causes (Jones 1969: 223–225). Nevertheless, what is significant is that, in the Aristotelian view, the acorn is seen to carry within itself the potential to become an oak tree, and the oak is the actualization of this inherent potential. The environment in which the acorn grows merely provides the nurturing medium for the tree which grows by actualizing the potential in the acorn. Thus the giant oak tree is seen by Aristotle to be the outcome of essential properties carried within the acorn, but not yet actualized in the world. Turning to a modern metaphor, Aristotle can be seen as thinking that the acorn carries a blueprint, program, or recipe for the oak tree as a formal cause. Only when this is combined with the final, material, and efficient causes does it produce an oak tree.

The Chinese organic view which explains phenomena by appeal to correlative causes is quite different. In this view the oak becomes what it is, not by virtue of some essence—the formal cause—within the acorn, but by virtue of its relation with other things within the context in which it grows. This does not deny that the acorn had a role, but this role is treated as only one factor among others to explain the growth of the oak. Hence, the Chinese explain the oak and its properties correlatively in terms of the larger context in which it grew, whereas Aristotelians explain it in terms of essential properties within the acorn. Both are causal explanations of the oak, but one looks to correlations with the context outside that treats the acorn as only a facilitating factor; but the other looks to causes within the acorn as essential, and treats the context as only a facilitating factor.[2]

The differences between the Aristotelian and Chinese organic views described above parallel the differences between those who see the genetic code as potentially carrying the form of the organism as a blueprint, program, or recipe, and those who see it as carrying recursive rules for cell multiplication and development that determine the form of the organism in dependence upon the context in which embryogenesis takes place. The Chinese view of correlative causation in which a thing grows in dependence on other things around it is closer to the latter view, which suggests that the correlative cosmology of Chinese natural philosophy may indeed be a response to properties in nature that grow in dependence on each other.

Chinese correlative cosmology also has an affinity with the central Buddhist notion that things arise in dependent origination on one another. It explains why Buddhism was received favorably in China, and also why the translation of Buddhist texts in China could draw heavily upon the terminology of Daoist thinkers closely associated with the organismic materialist view of nature.[3] This suggests, given that we have already found affinities between Buddhist and complementarity epistemologies, that Bohr may be right to see parallels in Daoist and complementarity epistemic views.

The affinity between Chinese correlative cosmology and the philosophical implications of the quantum theory was also recognized by the historian of Chinese science, Joseph Needham. He argues that quantum theory takes us beyond the mechanical vision of seventeenth century science and moves us closer to the correlative cosmology that characterized Chinese thought for millennia (Needham 1956: 582). Moreover, Needham also argues that among Chinese philosophical traditions it is the Daoists who most consistently emphasize the notion of nature as a correlatively conditioned self-regulating system of growing processes. He describes the Daoist conception of nature as follows:

> For the Daoists the Dao or Way was not the way of life within human society, but the way in which the universe worked; in other words the Order of Nature, which brought all things into existence and governs their very action, not so much by force as by a natural curvature in space and time, that reminds us of the Logos of Heraclitus of Ephesus, controlling the orderly process of change … the Dao was thought of not only as vaguely informing all things, but as being the naturalness, the very structure of particular and individual things. (Needham 1956: 36–37)

According to Needham the Daoists were concerned with the way of nature that lay outside the way of life in human society. This may be seen to follow from the Chinese organic materialist conception that things in nature grow and develop in correlative dependence upon others things without human intervention.[4]

This organic correlative vision of nature is central to the Daoist conception of how we should study and relate to nature. This is most clearly expounded by Laozi in his seminal text *Dao De Jing*.[5] It led him to recommend that we can only learn about nature by entering and communing with it without intervening in its processes. It is important to note that his desire for communion is not merely an expression of a secular wish to leave civilization—it is also connected with an urge to identify with nature—an identification so close and intimate that it is often seen as a sort of nature mysticism. Daoist mysticism, however, contrasts sharply with Hindu, Buddhist, Christian, and Islamic mystical traditions because it stresses communion with nature rather than withdrawal from nature. Its naturalistic orientation led the historian of Chinese philosophy, Fung Yu-Lan, to describe it as "the only system of mysticism the world has ever seen which was not profoundly antiscientific."[6]

The yin-yang complementarity principle is central to Daoist thought. In his D*ao De Jing*, Laozi appealed to the distinction between the *yin* and *yang* modes of action to articulate his approach to nature. *Yin* action, inspired by the feminine principle, is seen by him as yielding, non-dominating, and non-interfering action. By contrast *yang* action, inspired by the masculine principle, is assertive, dominating, and controlling. Laozi used it to articulate his opposition to the agricultural way of life emerging in China during the era of anarchy associated with what has come to be described as the

Warring States Period in Chinese history. The new agriculture of intensive cultivation required controlling nature by using sophisticated mechanical technologies for forest clearance and irrigation. Laozi saw these changes as promoting excessive interference with nature's spontaneity and inherent creativity. It led Laozi to call for a return to what Needham labels "the Order of Nature" and promote the notion that human society can flourish only by following the way of nature, which would let natural processes develop spontaneously. Laozi saw the method of intensive agriculture as driven by the masculine (*yang*) principle of controlling nature by violating its spontaneous growth, and saw his approach of yielding to nature's self-regulating processes as motivated by the feminine (*yin*) principle.

By raising the question of whether humans should adopt the masculine principle and control natural processes, or follow the feminine principle and leave them to develop spontaneously, Laozi opened the door to broader philosophical concerns. It led him to ask whether human nature, which shaped all social, political, and economic institutions should also be controlled or be allowed to express itself spontaneously. Clearly reorganizing wild nature into the cultivated system of agricultural society requires a parallel restructuring of human psychological and social identities. Hence, his call to follow nature's way without interfering with it also led to a wider debate in the ancient Chinese world, between Daoists and other schools of Chinese philosophy, about whether human nature too should be controlled or be left to develop without interference.

In contrast to all other major schools of Chinese thought in his time—the Mohists, Confucians, and Legalists—Laozi idealizes and defends the spontaneous way of life of the Neolithic culture of hunting, gathering and primitive cultivation that preceded the agricultural revolution. This culture largely depended upon processes of spontaneous regeneration, and succession in natural ecosystems, to produce food and other resources for human consumption. However, it was precisely this way of life that other schools of Chinese philosophy, especially Confucianism, sought to change. The alternative way they proposed involved agricultural intensification that not only required the cultivation of nature, but also the cultivation of human nature—in short, to create a new mode of production and new social institutions to facilitate it.[7]

Laozi's call for a return to a way of life that depended less on following the masculine principle and cultivating nature, and more on following the feminine principle and yielding to nature's spontaneous powers, came from his perception that the agricultural revolution led to social and economic dislocations on a vast scale. Many tribes and communities were forced to turn away from their earlier life of hunting, fishing, gathering, and subsistence farming.[8] He portrays the resulting hardships graphically in his *Dao De Jing*.[9] It leads to the intensification of competition, greed, and robbery [Ch. 3]; it sundered family relationships and created social turmoil [Ch. 18]; it inflamed lavish human desires that produced wars and famines [Ch. 46]; it proliferated prohibitions and laws [Ch. 74]; it promoted civil disorders and rendered people difficult to rule [Ch. 75]; it permitted the excessive consumption of a few supported by exploiting many who could barely feed themselves [Ch. 75]. There can hardly be a more vivid description of the traumatic and painful conditions in the Chinese world as people came to be forcibly co-opted into the emerging new agricultural order during the Warring States Period in which Laozi lived.

Laozi traces these social problems to the pervasive attempt to cultivate physical and human nature to serve the agricultural revolution. He questioned the assertive ideology of cultivating nature to increase productivity, which came to be adopted by the Mohists, Confucians, and Legalists. By doing so he rejected the underlying masculine dominative ethos of the agricultural revolution, and held up the feminine ideal of yielding to nature by letting it develop spontaneously. Only thus, he maintained, would harmony prevail in the world.

The *Dao De Jing* gives us a picture of Laozi's ideal society. In his social vision, people are seen to live best in relatively isolated self-sufficient communities with small populations; they do not use heavy tools, ships, carriages, or weapons; there is no writing; and people are satisfied with simple food, clothing, and houses without luxuries. In his model society, the rulers reign without formal laws, and there is no perceived need for sages to separate what is good and what is evil; or to teach rulers how to rule. [Ch. 80][10]

Thus the Daoist response to the crisis generated by the agricultural revolution was to call for a return to an earlier way of life modeled upon Neolithic cultures that preceded the agricultural revolution. This call for the restoration of ancient disappearing traditions also meant preserving the existing way of life for many communities yet to be drawn into intensive agriculture. It led Laozi to articulate epistemological, ethical, and political doctrines based on the feminine principle that contrasted sharply with those espoused by Confucian, Mohist, and Legalist schools of philosophy. Laozi's doctrines were designed to question, rather than consolidate, the dominative masculine orientation driving the agricultural revolution he saw as ravaging the world.[11] The issue between Daoists and others, especially Confucians, concerning whether nature should be cultivated by adopting the *yang* orientation, or be left alone by embracing the *yin* approach, became the central axis of philosophical debate in China. It continued to dominate debate until the dawn of the modern era.

Yet as the agricultural order consolidated itself in the Chinese world, the naturalistic concerns of Laozi became gradually invisible, and came to be displaced, even among the Daoists who followed him, by humanist concerns, so that Daoist debates with the Confucians became increasingly focused on social and psychological issues. The controversies became centered not on whether nature should be cultivated, but on whether human nature should be cultivated. The entry of Buddhist philosophy into China also made later Daoist thinkers interpret their views in psychological terms, so that Laozi's original naturalistic vision of a self-regulating spontaneous material nature became obscured.

Indeed, one could argue that Chinese philosophy itself originally developed as a debate between proponents of different ways of life—the Daoists representing the primarily hunter-gatherer and subsistence agricultural Neolithic tradition; the Confucians the agricultural tradition, the Legalists the nomadic way of life, and the Mohists the life of industrialists and craftsmen. This approach has some affinities with Chad Hansen's Daoist interpretation of Chinese thought in which he sees Daoism as a philosophically skeptical and reflective critique of the ethical debate between Mohism and Confucianism. However, it must be added, the approach here is that this ethical dispute is itself concerned with what sort of relations we ought to establish with nature that came to be obscured as Chinese civilization developed.[12]

This change was perhaps inevitable. Over time, the Daoists who came to address such issues found themselves increasingly surrounded by the cultivated nature of the agricultural order and more isolated from spontaneous wild nature. It became more difficult for them to identify with nature's self-regulation and the naturalistic orientation of Laozi became not only harder to recognize but also less relevant to their concerns. At the same time, the humanist concerns of the Confucians and the psycho-spiritual orientation of Buddhist philosophers, also put pressure on Daoists to reinterpret their doctrines by stressing the social, psychological, and spiritual concerns of the highly developed agricultural society in China.

Nevertheless, in order to explain the epistemological parallels between Daoism and quantum physics noticed by Bohr we have to return to the original naturalistic orientation of Laozi. We have to reread Laozi's epistemological proposals in the light of the issues that concerned him in his time. Let us begin our naturalistic interpretation by considering his well-known call to "act naturally". How should this recommendation to act naturally be interpreted? According to comparative philosopher Randall Peerenboom we cannot interpret the call to act naturally as requiring us to make our acts identical with acts of nature, since this would render Laozi's call either trivial or false. To support his conclusion Peerenboom argues as follows. If humans are a part of nature then all human actions are also a part of nature, so that whatever humans do will have to be taken as natural. Then the call to "act naturally" asks us to conform to a tautology and, since we cannot do otherwise, the call is trivial. But if humans are not a part of nature then whatever they do cannot be a natural action. The demand to act naturally would then require us to conform to an impossible injunction—one that is analytically false (Peerenboom 1991: 5–6).

Clearly, setting humans within nature makes their acts natural, whatever they do, and setting them apart from nature makes all their acts unnatural. This makes the call to act naturally either unnecessary or impossible. However, this is a false dilemma for there is third option: humans can be seen as a part of nature, but standing apart from non-human parts of nature. This would make it possible to interpret the injunction to "act naturally" as recommending that human actions should not interfere with the actions of non-human parts of nature. This would come about only if humans relinquish the masculine principle of controlling nature, and follow the feminine principle of yielding to nature's self-regulating ways. The scholar of Chinese philosophy and religion, Wing-tsit Chan, supports this interpretation of the Daoist dictum to "act naturally":

> The philosophy of Lao Tzu is not for the hermit, but for the sage ruler, who does not desert the world but rules it by non-interference. Daoism is therefore not a philosophy of withdrawal. Man is to follow nature but in doing so he is not eliminated; instead his nature is fulfilled.[13]

Chan interprets Laozi as making a distinction between the human and the non-human worlds, and maintaining that human nature becomes fulfilled only by allowing Nature to follow its natural spontaneous way.

Peerenboom rejects Chan's interpretation. He argues that it restricts the scope of the applicability of *Dao*, or Way of the Universe, since humans have the option of not conforming to its demand, thereby making it a principle that regulates only the non-human world. According to Peerenboom this limitation of the *Dao* is unacceptable because, he claims, *Dao* is a metaphysical principle that regulates all natural things including humans. But allowing humans to be capable of violating Dao would imply it cannot be "the all encompassing metaphysical principle of the universe" (Peerenboom 1991: 5).

Peerenboom's conclusion is questionable. Even though other schools of Chinese thought interpret Dao as a universal principle, it does not seem to be the view of Laozi, because it does not accord with his opening line in the *Dao De Jing*:

> "The *Dao* that can be told of is not the eternal *Dao*."[14]

It is clear that from the beginning Laozi makes a distinction between two kinds of *Dao* (Way), what he refers to as "the Way that can be told of" and "the eternal Way", and emphasizes that one cannot be identified with the other. It follows that Laozi does not treat the *Dao* as a universal metaphysical principle. Were this so then he would not refer to two kinds of *Dao*—one describable in words, and the other beyond words.

It is far more illuminating to read Laozi as propounding a practical doctrine promoting respect for the self-regulating way of nature and resisting the way of cultivating nature, which was becoming widespread at the time of China's agricultural revolution. This would give Laozi's opening line a clear-cut interpretation: it asserts that the way of cultivation that can be told of, or taught in words, is not the eternal way of self-regulating nature. In uncultivated wild nature, things grow spontaneously by following a timeless path—one which gets compromised when human beings impose the cultivated way of agriculture. Indeed the first line of the *Dao De Jing* points to the basic vision that informs the rest of Laozi's work—namely, that the eternal way of nature is not the way of cultivation. It defines his opposition to those teaching the new way of transforming wild nature into cultivated nature by adopting, without any restraint, the masculine principle of controlling natural processes. By setting himself up against the way of the cultivators, and asking people to emulate and follow the ancient and eternal untaught way of nature which preceded it, Laozi makes an appeal to the feminine principle of respecting self-transforming and self-regulating nature, which produces things by means both hidden and spontaneous.

The second chapter of the *Dao De Jing* confirms the interpretation we are recommending by continuing to develop the contrast between the eternal way and the way of cultivation[15]:

> Therefore the sage manages affairs without action (*wu-wei*)
> And spreads doctrine without words.
> All things arise, and he does not turn away from them.
> He produces them but he does not take possession of them.

In this passage Laozi appears to refer to the productivity of nature that produces without any contribution in the form of human action or knowledge. Moreover, unlike agricultural produce that can be appropriated by the cultivator, no one can claim the right to possess what nature spontaneously produces.

The passage also gives the ineffability of the eternal Dao a non-metaphysical and non-mystical meaning. The Dao is ineffable not because it is a transcendent universal principle or the eternal ground of existence, but because nature produces things by means of processes that are veiled from us.[16] Being mysterious to us, they elude description. Nevertheless, our lack of knowledge of these processes does not affect the birth and growth of things in nature since, as the passage emphasizes, "They grow by themselves". By contrast, the way of cultivating nature has to be taught, and be made manifest and transparent, by being put into words for those who cultivate things. Thus, one could suspect that the second chapter of the *Dao De Jing* carries Laozi's implicit indictment of the numerous agricultural manuals produced in China at that time—manuals considered crucial by the warlords of the various warring states concerned with increasing food surplus to feed armies needed both for their survival and expansionist ambitions.

This chapter of the *Dao De Jing* also introduces Laozi's notion of *wu-wei* whose significance for his philosophy is second only to that of Dao.[17] *Wu-wei* has been interpreted in diverse ways including 'doing nothing', 'performing action that yields to nature', or 'performing action that is natural'.[18] However, these prescriptions themselves may be applied in a natural, psychological, social, or spiritual context. We will find that if we ignore Laozi's naturalistic orientation, and construe *wu-wei*, in a psychological, sociological, or spiritual sense, then we would be led to conclude that Laozi is offering us a paradoxical notion. We are also likely to end up seeing *wu-wei* as carrying some inexpressible or mystical meaning.

This is exactly what happens when the comparative philosopher David Loy interprets the notion of *wu-wei* in his book, *Nonduality: A Study in Comparative Philosophy.*

Loy considers a variety of non-natural interpretations of *wu-wei*, and identifies problems with all of them only to ultimately offer a mystical interpretation as a way to resolve the paradox. One of the first non-natural interpretations Loy examines is the political interpretation of *wu-wei* as 'doing nothing.' On this interpretation, "doing nothing" can be taken to counsel a sovereign, a ruler, or a government, to essentially adopt a hands-off policy in managing affairs of state, based on the assumption that if people are left alone to lead their lives then social problems will not arise, and even if they did such problems would get spontaneously resolved. Loy argues that Laozi could not have intended this anarchist political approach since it would have failed miserably in "the cut-throat Warring States Period" in which he lived. Hence Loy rejects the political interpretation of *wu-wei* (Loy 1988: 99).

Loy next considers the personal interpretation of *wu-wei* as doing nothing. This reading was favored by some modern interpreters, such as Fung Yu-Lan, who argued that those who taught Daoism considered that "lying down is better than walking" (Fung Yu-Lan 1966: 255). Loy rejects this as the meaning Laozi intended when he used the term *wu-wei* because, he thinks, its call for no action whatsoever would not have been popular, or have resolved the problems Laozi faced in his time.

The third non-natural interpretation Loy considers is the popular psychological one. It suggests that practicing *wu-wei* involves performing action that is psychologically natural. However, what is meant by psychologically natural action can be construed in different ways. The fourth century Daoist philosopher Wang Bi treats it as action that involves no striving, but Fung Yu-Lan considers it to be action done without arbitrary effort. Loy criticizes both these interpretations. He argues that unless we provide adequate criteria to demarcate actions done with striving from those done without, or distinguish arbitrary from non-arbitrary effort—specify criteria to separate willful and non-willful actions—calls to act in psychologically natural ways are meaningless (Loy 1988: 101). Moreover, even from a psychological point of view there is no coherent way to distinguish willful action from other kinds of action. Consequently, Loy concludes that interpreting the call to act naturally in psychological terms is ultimately vacuous.

Not only does Loy dispose of the political, personal, and psychological interpretations, but he also raises objections against naturalistic interpretations of *wu-wei*. He considers the popular interpretation of *wu-wei* as referring to action that does not employ force but, instead, yields to forces from without when such yielding action may sometimes be more effective than resistance. For example, pine branches which are unyielding often break under the weight of a snowfall, but the branches of the willow which bend under the burden are able to drop accumulated snow and spring up again. Laozi makes the same point by referring to water which always manages to get to its final destination because, being soft and yielding, it bends around obstacles rather than trying to remove them (Loy 1988: 100).[19]

The above notion of yielding to forces without could be given a slightly different construal—we may interpret *wu-wei* to convey the idea that a slight action, taken at the right time, can be made to accomplish extraordinary results. Yielding action is one that does not resist the self-regulating powers of nature but works with them. Loy refers to the example of a sapling whose growth can be influenced to a far greater extent than that of a mature tree (Loy 1988: 100). The sapling example is particularly apt because one uses the processes that drive the growing plant to produce extraordinary outcomes by taking a small action at the beginning—bending a branch here, pruning one there—that produces large effects on the mature tree.

Although Loy does not see the naturalistic interpretation of *wu-wei* as necessarily impractical, incoherent, or vacuous, in the way he takes the political, personal, and psychological interpretations to be, he nevertheless considers it inadequate. He thinks it cannot accommodate some of the wider epistemological and metaphysical claims generally associated with Daoism. In particular he thinks that it does not make sense of the so-called "Daoist paradoxes"—the action of non-action, the knowledge of non-knowledge, and the morality of non-morality. Many of Laozi's recommendations in the *Dao De Jing* seem to suggest that action, knowledge, and morality are somehow intimately bound with their opposites. Loy thinks that such a unity of opposites can only be accommodated by giving *wu-wei* an interpretation that transcends the duality of an acting subject and the action performed. For Loy the paradoxes refer to, what he terms, 'nondual actions'—actions that involve "no bifurcation between subject and object" (Loy 1988: 96). He thinks that Daoist thinkers, like Indian Advaita Vedanta and Japanese Zen Buddhist philosophers, use such paradoxes to point toward a metaphysical non-dualism whose final goal is the achievement of a mystical experience that transcends the subject–object dichotomy. Thus, although Loy thinks that a naturalistic interpretation of *wu-wei* is better than a political, personal, or social interpretation, he nevertheless considers that we need a mystical interpretation to deal with the Daoist paradoxes because we cannot make sense of them in naturalistic terms.

Loy's mystical interpretation of *wu-wei* is questionable, although this interpretation did become popular among Daoists following the arrival of Buddhism, simply because the naturalistic interpretation became more difficult to accommodate as the agricultural revolution consolidated within China. But Loy is wrong when he supposes that the so-called Daoist paradoxes refute all naturalistic interpretations. Indeed, not only can these paradoxes be given such an interpretation but doing so brings out even more clearly Laozi's call to defend the way of nature against the way of the cultivators.

On the naturalistic interpretation the distinction is between nature's natural actions and humanly engineered non-natural actions. Then *wu-wei* involves humans following the feminine non-assertive path of non-action in order to allow nature to follow its own path of environmentally regulated action. The resulting actions brought about by nature without human intervention can now be seen as done without striving and arbitrary effort on our part.

Such an approach would be endorsed by many environmentalists and ecologists today, and explains the popularity of Laozi in such circles. Laozi is seen as recommending that we protect nature by doing nothing to interfere with her processes—that is, we preserve wilderness areas by leaving them alone, or safeguard wild-life habitats by proscribing human interference. Indeed doing nothing in such situations, or taking no action, is probably an excellent strategy for conserving wildlife and its habitats. It also accords with the deep social consciousness that originally informed the *Dao De Jing* for the sufferings of Neolithic hunter-gatherers and subsistence farmers drawn into the agricultural order.

To see the viability of the naturalistic approach to the Daoist paradoxes let us begin by looking at how Loy approaches them in his mystical interpretation.[20] Consider the paradox of the action of non-action that Loy describes as central to Daoism (Loy 1988: 97). It seems to be violating logic or semantics by asserting that non-action, the denial of action, can also be action. Taken at face value, one may be tempted to either deny outright that it expresses anything meaningful, or seek some mystical sense that can render coherent the *prima facie* contradiction it presents. Loy describes the epistemological quandary it raises as follows:

> The main problem with understanding *wei-wu-wei* (the action of non-action) is that it is a genuine paradox: the union of two contradictory concepts, nonaction ("nothing is done…") and action ("…and nothing remains undone."). The resolution of this paradox must somehow combine both, but how this can be anything other than a contradiction in terms is difficult to understand. (Loy 1988: 101)

Nevertheless, we have already confronted similar paradoxes with quantum theory and Indian epistemologies. In both cases paradoxes were seen when something gets perceived differently through two mutually exclusive contexts. Thus a quantum object may appear to be a particle or a wave when perceived through different contexts. Similarly a thought-form (gestalt) may appear to be given to us as spectators or constructed by us as actors in different contexts. We could even say that a cell in a growing frog embryo may be seen as an eye cell or a muscle cell depending on the context in which it is located. In all these cases ignoring the context of observation generates paradoxes that appear to point to contradictions. However, what they really show is the need for adopting complementary viewpoints—viewpoints that are mutually exclusive, but equally necessary—to fully comprehend the phenomena concerned whether these be about quantum, gestalt, or biological properties. We will find that the Daoist paradoxes also point to a similar need to adopt complementary perspectives so that judgments we make are seen as shaped by mutually exclusive, but equally necessary, contexts.

Returning to the paradox of the action of non-action, it is possible to see the terms action and non-action as being used in two different contexts—the natural context and the human context. The term 'action' refers to the spontaneous activity of nature that takes place without human intervention, that is, when humans engage in no action. It is precisely human non-action, or human non-interference in natural processes, that facilitates nature's self-regulating action. Nature acts on its own provided humans do not disturb it; its action is made possible by human non-action. When humans adopt the feminine yielding approach, nature can assert its self-regulating powers. The action of non-action expresses the fact that nature's action is facilitated by human non-action.

Similarly the paradox of knowledge of non-knowledge can also be explained if we take into account that there are two contexts in which knowledge can be considered. First, we have human knowledge, or more precisely human know-how, which is embodied in our practical knowledge concerning how to go about cultivating nature. Second, we have the knowledge, or more exactly know-how, embodied in nature, and carried as information contained in natural systems that allows them to self-regulate themselves. The two kinds of knowledge are also what Laozi sees as carried by "the Dao that can be told of" and the "eternal Dao". Non-knowledge refers to our ignorance of nature's way, and knowledge is the recognition of this ignorance on our part. The knowledge of non-knowledge is the knowledge that we do not have knowledge of the way of nature.

The importance of being aware of our ignorance was also stressed by Confucius in a different context, when he argued that the wise man is one who not only knows what he knows, but also knows what he does not know. In the Hellenic world, Socrates also attempted to convince his contemporaries that in many areas of the moral sphere they did not really know what they thought they knew—he considered knowledge of such non-knowledge important. His goal was to emphasize the significance of a critical outlook that would recognize ignorance where it existed. Laozi stresses the same critical attitude with regard to the presumption that we know how nature works, by maintaining that often humans do not really know what nature 'knows'. Such lack of knowledge of our non-knowledge may lead us to intervene in nature inappropriately by obstructing its self-regulating way.

This interpretation of the paradox of the knowledge of non-knowledge would explain Laozi's otherwise strange opposition to learning[21]:

> The pursuit of learning is to increase day after day.
> The pursuit of Dao is to decrease day after day.
> It is to decrease and further decrease until one reaches the point of taking no action.
> No action is undertaken, and yet nothing is left undone.

In the first line, the term "learning" refers to accumulating positive knowledge of how to act. In the second line it involves relinquishing such positive accumulation, and learning to recognize one's ignorance of nature's way. In the process, though we do less and less, it is precisely by such non-action that we permit the free play of nature in which nothing, or rather nothing important, is left undone. In short, it is by not using our human knowledge to control nature that we allow nature's knowledge to express itself.

If the knowledge of non-knowledge refers to the knowledge that the know-how of nature operates without human intervention, then the morality of non-morality refers to the morality, or regulative principles, inherent within nature that operate when we do not interfere by imposing action motivated by human morality. The morality paradox arises from our use of the term morality in different contexts—nature's morality inherent in its self-regulating principles and human morality defined by our social norms. The morality of nature is followed only if we suspend the imposition of human norms that interfere with this. The paradox of the morality of non-morality can then be seen as directing us to adopt the feminine principle and yield to nature's morality by not asserting our morality over nature.

Such a (non-)moral standpoint leads Laozi to write:

> When the Great Dao declined, the doctrine of humanity and righteousness arose. [Ch. 18]

It also leads him to assert that ugliness and evil entered the world only when human beings imposed their principles upon nature:

> When the people of the world all know beauty as beauty,
> There arises recognition of ugliness
> When they all know the good as good
> There arises the recognition of evil. [Ch. 2]

The passage above is reminiscent of the biblical *Book of Genesis*, which sees suffering and pain enter the human world only after their knowledge of good and evil led Adam and Eve to be expelled from the Garden of Eden—a garden where all things needed were produced without toil and effort. It is possible to suppose that both Laozi and the biblical narrative of Eden are referring to humanity's primordial relations with self-regulating nature and the tumultuous transformation subsequently wrought by the agricultural revolution.[22]

Thus, despite their differences, the Daoist paradoxes, like the quantum and Buddhist paradoxes, are also designed to point to the context dependence of our judgments. Combining judgments made in two mutually exclusive contexts can often lead us to think that we are facing contradictory claims where none exist. The Daoist paradoxes are intended to reveal that terms like 'action', 'knowledge', and 'morality' can be construed in either the human or natural contexts. Unless we recognize the context dependence of terms like 'action', 'knowledge', and 'morality', since they can be applied with reference to human or natural contexts, we might be tempted to conclude that the Daoist paradoxes violate the rules of logic.

Indeed, if we treat nature as possessing no inherent knowledge, that is, as not containing natural information, or as bereft of spontaneous creative action, then all knowledge and action would have to be attributed to humans alone. Nature would then seem to be both *blind* and *dead*. One could argue that the mechanical view of the universe that emerged in the seventeenth century did make nature appear to be blind and dead by ignoring its self-regulating intelligence that directs growing things in nature. It leads us to mistakenly assume that knowledge, morality, and action are attributes of human agents alone. This would lead us to look at the Daoist paradoxes only from the point of view of human knowledge, morality, and action. We would then see talk of 'the knowledge of non-knowledge' as incoherent, and be led to ask how non-knowledge could be knowledge. Similarly, by ignoring the different contexts in which the terms morality and action get applied, we would be inclined to wonder how human morality could also be non-morality, or human action could be non-action. As a result, the so-called Daoist 'paradoxes' would appear to violate the rules of standard logic, especially the principles of non-contradiction and excluded middle.

However, the goal of the Daoist paradoxes is similar to the goal of the Buddhist and quantum paradoxes. They are intended to liberate us from acontextual patterns of thinking that are inappropriate in dealing with grown properties in nature. The paradoxes associated with quantum complementarity are a response to the way atomic properties grow in an experimental context we have selected; the paradoxes associated with the Madhyamika negative tetralemma are a response to the way we read sensory stimuli in dependence on a theoretical context. This theoretical context is itself a response to the fact that things are more easily and sharply identified if we take into account the environmental context in which we find them. Similarly the Daoist paradoxes emphasize the context dependence of growing things—the human context of agricultural production and the natural context of spontaneous production in untamed nature. What the Daoists commend is that we do not adopt the dominative masculine principle that leads humans to cultivate nature, but follow instead the yielding feminine principle that lets nature express its self-regulating powers. In short, they recommend that the *yin* approach is preferable to the *yang*.[23]

4.2 Liberating Gaia from Bacon: Transcending the Yang Principle

The complementarity of the masculine and feminine principles, or the cultural and natural, can also illuminate epistemological issues raised by ecofeminists. The ecofeminists, like the Daoists, appeal to the feminine principle as a way of redefining our relations with nature in the face of the current environmental crisis. They are critical of the Enlightenment ethos that continues to inform modern science and its epistemology which they perceive as driven by the masculine principle of dominating nature. They see this assertive patriarchal orientation of control as a threat to the processes that self-regulate growing things in nature. Many ecofeminists also go further by arguing that the problem is not simply rooted in the application of scientific knowledge, but in the content of current scientific theory and its methodology. They maintain that both need to be reformed if we are to prevent the systematic degradation and destruction of nature.

Their call has evoked a strong counter-response from the scientific community. Most of these critics of ecofeminism do not dispute feminist attempts to ensure gender equality by striving to secure more and equal places for women in the sciences, encourage greater focus within science on problems faced by women, or eliminate sexist language in science publications and journals. They do not even object to feminists criticizing the misuse of scientific knowledge to produce a technological and economic system that violates the earth's ecological integrity. But they take issue with the notion that there can be a legitimate feminist critique of either scientific theory or scientific method.

They charge that attempts by ecofeminists to show that scientific theory and method are deformed by the masculine principle are simply efforts to expose unfortunate sexist *metaphors* sometimes deployed by ill-informed scientists when they address theoretical and epistemological concerns in science. For example, in their widely read study, *Higher Superstition: The Academic Left and its Quarrels with Science*, Gross and Levitt complain that most feminist critiques of scientific theory and epistemology constitute nothing more than efforts to microscopically scrutinize the language of science and show it to be "tainted by sexist ideology" (Gross and Levitt 1994: 116). But, say Gross and Levitt, even if feminists rightly find fault with the masculine or anti-feminine metaphors sometimes used by scientists, they wrongly conclude that this renders questionable the theoretical and epistemological approach of science *per se*. Hence, Gross and Levitt conclude, since "metaphor mongering is the principal strategy of much feminist criticism of science" it does not present any serious challenge to the substantive *content* and *methodology* of science. Feminists must go beyond mere metaphor criticism, they contend, if they wish to make a convincing case against current science and its practices:

> We would have to be shown that there are palpable defects, due to the inadequacies of a male perspective, in heretofore solid-looking science and that the flawed theories can be repaired or replaced by feminist insights. The issue before us is knowledge, scientific knowledge specifically, and the extent to which the prevalent feminist critique, as agent of methodological or conceptual change, is relevant to its advance … [However] the feminist critique is overwhelmingly concerned with metaphor, rather than with the logical content and analysis of scientific results. But scientific results, we must insist, are not simply metaphors. (Gross and Levitt 1994: 112)

However, there is more to metaphor than meets either Gross' or Levitt's eye. This becomes evident when we examine closely feminist critiques of what has often been held as one of the key pillars of modern science—Francis Bacon's inductive and experimental method. In the process of formulating and describing his method Bacon frequently appeals to the metaphor of nature as a woman whose secrets can only be extracted by deploying inquisitorial techniques. This has been particularly noted by the feminist historian of science Carolyn Merchant.[24] She argues that the conception of nature as a female organism prevailed in Western culture for millennia leading to an attitude of both reverence and respect for nature. A change in this Western orientation came about after Bacon articulated his new mechanical conception of nature in which he saw nature as a female to be dominated by the male. Merchant demonstrates that Bacon's account of his experimental method defines the relationship of humans to nature in terms of a dominant male over a submissive female.[25] She quotes him speaking of following and hounding female nature in her wanderings so that "you will be able when you like to lead and drive her afterward to the same place again." Bacon advises the scientist to compel nature to reveal her secrets by "entering and penetrating into those holes and corners, when the inquisition of truth is his whole object".[26] He recommends observing nature not in her natural state, but when she is made to deviate from it by treating her as we do with mechanical artifacts:

> She is either free and follows her ordinary course of development as in the heavens, in the animal and vegetable creation, and in the general array of the universe; or she is driven out of her ordinary course by the perverseness, insolence, and forwardness of matter and violence of impediments, as in the case of monsters; or lastly, she is put in constraint, molded, and made as it were new by art and the hand of man; as in things artificial.[27]

But Gross and Levitt repudiate such feminist critiques of Bacon. They argue that his metaphors of domination and torture cannot influence the content of scientific theories or impugn his fundamental experimental methodology. Nothing is proven about the limits of modern science, they argue, by appealing to the "turgid metaphors of Bacon" (Gross and Levitt 1994: 123). Instead they propose a toast to Bacon as a methodologist:

> Let us raise a glass to Bacon! He wasn't much of a scientist or mathematician, but he made some shrewd guesses as to how our species might crawl out of the rut of ignorance. And here's to Bacon's science—if that misattribution is to persist in our universities—Baconian in the sense of a rigorous adherence to the empirical, and a faith that what we learn that way can improve the prospects for human life. The more Baconian science we get, the easier it will be to believe that we have a fighting chance, if no more than that, on this lovely planet that spins its way through an unimaginably violent—and indifferent—space. (Gross and Levitt 1994: 178)

Nevertheless, in spite of the resounding endorsement they give to Bacon's experimental method, Gross and Levitt also appear to underestimate Bacon's role in shaping modern science. He may not have been a great scientist or mathematician, but many great scientists idolized Bacon as the founder of the method they followed. Newton saw himself as adopting Bacon's method, and the Royal Society in London considered him the source of its founding inspiration. The *philosophes* of the Enlightenment in France saw in him a pioneer who discovered the inductive-experimental method. In 1830 the renowned scientist John Herschel gave him high praise in his influential work *Preliminary Discourse on Natural Philosophy*:

> By the discoveries of Copernicus, Kepler, and Galileo, the errors of the Aristotelian philosophy were effectually overturned on a plain appeal to the facts of nature; but it remained to show on broad and general principles, how and why Aristotle was in the wrong; to set in evidence the peculiar weakness of his method of philosophizing, and to substitute in its place a

> stronger and better. This important task was executed by Francis Bacon, Lord Verulam, who will, therefore, justly be looked upon in all future ages as the great reformer of philosophy, though his own actual contributions to the stock of physical truths were small. (Herschel 1851: 113–114)[28]

In order to appreciate the force of feminist arguments not only against Bacon's inductive experimental method, but also its influence on the content of scientific theories, let us begin with Bacon's famous dictum: "We can only command Nature by obeying her."[29] *Prima facie* it appears extremely paradoxical and reminiscent of the Daoist paradox of the action of non-action. How can one both command and obey Nature at the same time? The answer is that, like the Daoist philosophers, Bacon is exploiting contextual ambiguity by referring to two different contexts through which we relate to Nature.

These two contexts become evident when we put together Bacon's doctrine of the four idols—idols against which we ought to guard ourselves—with his method of controlled experiment. We will find that combining these two methodological orientations enables Bacon to offer an approach to nature that leads not only to the discovery of the universal laws that condition natural systems but also leads scientists to overlook the distinctive contexts that shape grown properties in nature. Thus, by convincing others to take the methodological approach he recommended, Bacon also influenced the content of the scientific theories that came to be discovered. His methodology produced systematic knowledge of nature's universal laws along with systematic neglect of natural contexts at the same time.

Let us begin by looking at Bacon's well-known proscriptions against what he called 'idols of the mind'. Take the first of his idols—the idols of the tribe which he declares:

> [...] lie deep in human nature itself and in the very tribe or race of mankind. For it is wrongly asserted that the human sense is the measure of things. It is rather the case that all our perceptions, both of our sense and of our minds, are reflections of man, not of the universe, and the human understanding is like an uneven mirror that cannot reflect truly the rays from objects but distorts and corrupts the nature of things by mingling its own nature with it.[30]

Two points made by Bacon are worth noting. First, he warns us against the proclivity of the human mind to project imagined attributes onto the scheme of things. Second, he appeals to the metaphor of the mind as a mirror that, when 'uneven', distorts the objects it perceives. Hence it is imperative that we polish the mirror of the mind so that our perceptions of the universe would faithfully reflect the object perceived. Bacon is clearly advocating humility in our approach to nature so that our perceptions would bear reference to the universe rather than to our minds.

Bacon then proceeds to delineate the other idols that might distort our understanding. His idols of the cave derive their origin from the peculiar nature of each individual's mind and body, and also from what an individual has "been taught or gained in conversation with others, or from his reading, and the authority of those whom he respects and admires." The idols of the market are those "arising from the dealings and associations of men with one another ... because of the commerce and meeting of men there." The idols of the theater are those "which have crept into human minds from the various dogmas of philosophies, and also from faulty laws of demonstrations." These include received or imagined systems of thought that are like "so many stage plays creating fictitious and imaginary worlds."[31]

Having delineated the ways in which the idols can mislead us Bacon goes on to recommend that we purge ourselves of all the idols. He recommends that we abjure and renounce them with resolution "so that the entry into the kingdom of man, which is founded on the sciences, may be like the entry into the kingdom of heaven, which is only to be entered as a little child."[32] Thus there is nothing in Bacon's warnings against seduction by the idols which would suggest a project to dominate nature. He seems rather to be proposing a kind of epistemological psychotherapy designed to purge the mind of distorting influences, prejudices, and pre-judgments so that it would become receptive to nature.

Bacon's method is clearly designed to repudiate what he sees as a general phenomenon—the tendency to impose theoretical interpretations on information obtained through the senses. Bacon does not see such theoretical interventions as a process of taking into account contextual information that enhances our capacity to read sensory stimuli. He does not view theories as instruments of perception that amplify the power of our senses. Rather he sees them as distorting the capacity of the senses to mirror nature. The doctrine of idols in Bacon's methodology is thus designed to diminish our sensitivity to natural contexts. It recommends viewing a thing independent of our theoretical understanding of the way things interact and influence each other. We have already seen how theoretical understanding can transform our theories into instruments of observation that amplify our sensibilities. Moreover, *contra* what Bacon assumes, the role of theories as observational instruments does not impugn the objectivity of scientific knowledge—in fact it enhances objectivity by imposing additional constraints on the construction of theories, since theories that serve well as instruments of inference may yet fail as instruments of observation. Hence, Bacon's repression of the role of theories in observation by appeal to his doctrine of idols weakens the objectivity of scientific knowledge. Moreover, since it is not possible to purge our minds from the influence of idols altogether, he further weakens our objectivity by ideologically precluding awareness of the theoretical projections that shape the observations that lead to our inductive conclusions.

Bacon's repression of contextual knowledge promoted by his doctrine of the idols becomes reinforced by his proposed experimental method. Indeed, the repression of contextual knowledge that results from the application of his experimental method lies at the root of the feminist critique of his method. Bacon's experimental method recommends adopting techniques that are designed to *compel* nature to reveal the underlying laws that regulate her behavior. Merely observing phenomena in their natural contexts, he argues, cannot allow us to discover these laws of nature:

> A natural history compiled for its own sake is quite unlike one collected in an organized way with the aim of informing the intellect and building a philosophy. And these two [kinds of] histories, different as they are in other matters, differ especially in this, that the former contains only the variety of natural species and no experiments of the mechanical arts. And just as in ordinary life the true personality of a person and his hidden thoughts and motives show themselves more clearly when he is under stress than at other times, so things in Nature that are hidden reveal themselves more readily under the vexations of art than when they follow their own course. There will be therefore be grounds for optimism regarding natural philosophy when, and only when, natural history (which is its basis and foundation) shall have been better organized; but until that is done, hardly any.[33]

The above passage summarizes Bacon's experimental method, which involves studying the behavior of things not within their natural contexts, but within the context of controlled situations set up by the scientific observer. Bacon argues that natural history itself—the study of plants and animals—must adopt the experimental method so successful in the mechanical arts in order to make progress. Hence, Bacon complements the insensitivity to contexts in perception promoted by his epistemological psychotherapy of purging the influence of the idols, with his more aggressive experimental approach in which natural contexts are displaced by contexts strictly set up by the observer.

Can we say that, despite his repression of contextual sensitivity in the act of perception and experimentation, Bacon's epistemological psychotherapy, followed by controlled experimentation, really does promote complete knowledge of natural processes? It may create the illusion that it does for two reasons. First, his doctrine of idols seems to exclude contamination by ancestral, cultural, personal, and philosophical projections by freeing the mind from prejudices. Second, his experimental method seems to make possible the identification of the universal laws that constrain nature under diverse and varied contexts devised by scientists. However, the experimental method also reveals nature's laws only after we adamantly violate the natural contexts in which these laws normally operate Indeed, Bacon suggests that these laws can only be discovered by setting up artificial experimental contexts not found in nature, since "the secrets of nature reveal themselves more readily under the vexations of art … than when they go their own way."

We can now understand Bacon's apparently paradoxical dictum "Nature to be commanded must be obeyed." What he is recommending is that once we come to know nature's inviolable laws to which we have to conform, we will have greater control over nature. Although we cannot change natural laws we can freely alter natural contexts, as we do in the experimental method, and set up new contexts defined by us in order to make nature serve us. What we are able to command are the contexts of operation of natural laws; what we are forced to obey are the natural laws themselves.

Bacon's science is really concerned only with the discovery of inviolable, invariable, universal, fundamental, and general laws of nature that continue to operate in all the artificial experimental contexts that the scientist can imaginatively devise. Contextual information is treated as a distorting force in the act of observation by his doctrine of idols and, as of secondary significance, in his experimental method.[34] Bacon is recommending that we attempt to discover acontextual general laws by experimental scrutiny that alters nature's context that is, 'vexes nature,' to use Bacon's expression. The experimental method thereby separates two different elements in nature. First, there are the universal laws over which the scientist has no control and which constitute the goal of scientific inquiry. Second, there are the various contexts within which these laws operate. These contexts can be, and have to be, controlled and varied by the experimenter in order to identify the laws that remain unchanged under all alterations. It is by controlling natural contexts that we discover natural laws—laws that limit our capacity to manipulate nature. Hence, Bacon's dictum that in order to command nature we must obey her is not a paradox that violates the rules of logic—it does not ask us to command and obey the same thing. Rather it suggests that in order to command natural contexts we must obey natural laws.

Moreover, his inductive and experimental method teaches scientists to violate nature's contexts in order to discover her secret laws. This could easily lead science to ignore nature's context altogether—especially if such contexts are complex, inaccessible, and not directly perceivable. Such an orientation is neither likely to promote respect for natural contexts, nor reverence for nature's contextual integrity. Yet Bacon's methodology ignores the crucial fact that it is precisely complex natural contexts—often little understood by human beings—that makes possible the diversity of living things in the world. What makes different things unique in the world and shapes their specific behaviors, so that they are distinct from each other, are not the universal laws to which all are subject. Their uniqueness arises from the different contexts in which these laws function. Butterflies and bears, lemurs and lions all conform to the laws of physics, but what makes each so different are the genetic and environmental contexts in which these physical laws operate to generate their unique and different forms and behaviors.

By emphasizing the discovery of general laws, and treating the contexts in which they function as of secondary cognitive significance, Baconian science distorts our understanding of natural systems and promotes insensitivity to contextual environmental knowledge. He does this not only by deluding us into thinking that his epistemological psychotherapy can purge our minds from the distortions of contextual knowledge, but also by making us adopt an experimental method that violates natural contexts in order to produce knowledge. It is precisely this repression of contextual knowledge that ecofeminists reject. They argue that Bacon's metaphors are far from innocent decorations to an otherwise acceptable method—his metaphors generate a method that valorizes knowledge of universal laws and discounts local contextual knowledge as secondary. But ecofeminists argue that we need to pay regard to nature's context to promote living in harmony with natural systems, and preclude the systematic violation of natural systems that the Baconian ethos promotes. Respect for natural contexts that make possible the self-regulating powers of nature is what ecofeminists demand when they call for a receptive, non-dominating, non-exploitative, integrative, cooperative, and caring relationship with nature. This is also at the heart of the Daoist appeal to the feminine principle when they articulate their epistemological critique of agricultural civilization.

The importance of making the scientific method sensitive to contextual knowledge has also been stressed by the philosopher of science Stephen Toulmin. He argues that Enlightenment science, in general, repressed contextual knowledge, and describes the change he wishes to promote to transcend the limits of the modernist vision initiated by Bacon and Descartes as follows:

> [O]ne aim of 17th century philosophers was to frame all their questions in terms that rendered them independent of context; while our own procedure will be the opposite—to recontextualize the questions that these philosophers took the most pride in decontextualizing. (Toulmin 1990: 21)

Toulmin rightly argues that Enlightenment science shifted attention away from contextual knowledge valued in medieval European and other cultures, inspired by an organic world view, to acontextual knowledge of universal laws valued by early modern thinkers. According to Toulmin the emerging science of chaos and complexity, by recognizing the sensitivity of complex system behaviors to their initial conditions and stressing the resultant impossibility of predicting the behavior of such systems even after we have knowledge of the natural laws they obey, is once more revealing the importance of contextual knowledge for science. Consequently chaos theory, along with quantum physics, gestalt psychology, and ecology, can be taken as shifting us away from the acontextual tradition of early modern mechanical science. This makes the feminist and Daoist emphasis on the importance of contextual knowledge obtained by communing with natural systems, rather than experimenting on them, an indispensable corrective to the limitations of Baconian science and its methodology.

However, such a defense of ecofeminist critics of Baconian science might not satisfy everyone—especially those who remain convinced that these critiques are not significant because they have not made any noteworthy contributions to advancing the content of a scientific theory. Consider Gross and Levitt on this matter:

> The central argument varies from one critique to the next, depending upon which of the sundry standpoints within feminism the critic represents. Nevertheless there are broad agreements. The firmest of these is that feminist insight and practice must, by definition, improve the range and depth of scientific theory, and must by definition eliminate errors arising from unconscious commitments to patriarchal assumptions. Thereby, the validity of science, as well as its scope, are to be enlarged. On the other hand the influence of postmodernist theorizing is not absent: many feminist tracts accept and defend the notion that there is no "objective" science, merely a variety of "perspectives," one of which—patriarchal science—has been "valorized" and "empowered" so as to preclude until now the possibility of a feminist science. On occasion, finally, feminism joins hands with New Age attitudinizing, yearning for the rebirth of a prelapsarian golden age, wherein the human race knew and worshipped a goddess-nature, without artificial categories, tortuous celebration, and elaborate physical devices of male technology. (Gross and Levitt 1994: 109)

In the above passage Gross and Levitt completely miss the point of ecofeminist criticism. They are still concerned with scientific theory—not with the contextual situations in which theories function. They cannot comprehend the feminist critique because they do not have any appreciation of the significance of having 'knowledge of *their* non-knowledge' of natural contexts—such knowledge of their ignorance is needed to enable them to see 'the action of *their* non-action' in natural systems. The issue is not whether ecofeminists can propose another scientific theory. The issue is whether contextual knowledge of how things grow in nature is important—both in our acts of observing nature and our interactions with nature.

Only by acknowledging the importance of contextual knowledge can we understand why ecofeminists reject patriarchal orientations that deploy tortuous technology violating natural contexts; and why they 'worship' a female nature Gaia who was better served in a "prelapsarian golden age" when there was reverence for her diverse contexts. In their different ways both Gaian ecofeminists and feminist critics of Bacon are attempting to once again valorize contextual knowledge of nature as well as respect for natural contexts. The issue is not the formulation of new universal theories that can replace current theories. The issue is whether we can transform scientific thinking by going beyond the unrestrained pursuit of dominating natural contexts by total subservience to the masculine principle and by allowing the feminine principle of yielding to natural contexts once more to inform science.

Moreover, we can also take issue with the charge that feminists reject "objective" knowledge by embracing cognitive perspectivism. Not many feminists are as relativistic as Gross and Levitt suppose. Were this so, even the feminist critique of Baconian science would become untenable. Most feminists would want to argue that a one-sided emphasis on patriarchal values of domination, without the corrective complement of feminine values that respect nature's contexts, leads to poor science. They want to suggest that such a science is not objective enough—it valorizes knowledge of nature's acontextual objective laws at the expense of objective knowledge of the natural contexts in which these laws operate. Hence, ecofeminists would say that Bacon should have developed one step further his idea that 'in order to command nature we should obey her', by adopting the principle of yielding not only to nature's acontextual laws, but also nature's contexts.

Let us now consider how our current approach to agriculture would get transformed if we yield to natural contexts by adopting the feminine principle endorsed by both ecofeminists and Daoists. Respecting nature's contexts would lead us to appreciate the Daoist epistemological call to have knowledge of our non-knowledge of natural processes, follow the morality of non-morality that respects nature's way, and pursue the action of non-action that yields to nature's action. Such sensitivity to natural contexts would give nature's self-regulating powers far more scope to express themselves. However, the idea that agricultural production could be managed by self-regulating technologies may seem strange, since it is often assumed that human progress came about precisely by replacing wild nature by a cultivated nature using mechanical technologies. Today's agroindustry has gone even further by recasting the very image of nature in a mechanical metaphor.

Indeed agriculture has become so closely linked with the process of replacing natural self-regulating technologies with human technology that the very concept of returning to natural technologies as recommended by Daoists and ecofeminists appears a violation of agricultural practice. In order to see how agricultural practices can be more sensitive to natural technologies let us examine more closely three models of agriculture. The first model is the agro-industrial system we have adopted today; the second is the model of agriculture in the new civilization of cultivation that the Daoists opposed; and the third is the permaculture practiced by Neolithic cultures, which the Daoists recommended emulating.

Consider modern agriculture. Inspired by a mechanical vision of nature it operates more on the lines of a factory assembling machine parts than an organic system sensitive to the cycles of nature. The contrast to how plants grow in a self-regulating forest system cannot be more striking. In the forest there are ecological cycles that ensure the movement of vital substances from locations where they are present to those where they are needed. Phosphorous, carbon, and nitrogen are important chemical elements that participate in such cycles. Moreover, in the case of nitrogen, living organisms like bacteria and a few blue-green algae are involved in 'fixing' it in a form amenable to assimilation by plants. The nitrogen taken up by plants is absorbed by animals that consume them, and is finally returned to the soil through their feces and decomposing bodies to be reused by plants.

Moreover, weeds with deep roots play a role in the ecological balance by drawing vital trace minerals from the depths of the soil to the surface; and earthworms play an important role in breaking up and loosening the soil so that it can better retain moisture and absorb vital nutrients from decomposing wastes. The soil itself is often built up by the material of dead plants and organisms that enrich and shape it into a more fertile medium for new generations of growth. This is an extremely complex process, which is ultimately driven by solar energy and little understood even now. The energy is trapped by plants through photosynthesis, which itself is part of a cyclical process. Plants absorb carbon dioxide and release oxygen; the oxygen is absorbed by animals for their energy needs with resultant release of carbon dioxide.

Modern agricultural systems do not operate on such ecological cycles. By making agricultural lands support distant urban centers with a vast appetite for food, the nutrients from the ground are removed to faraway places and flushed down sewage systems which do not return them to the original farmland. In order to replace this loss, modern agriculture has to employ fertilizers that draw upon fossil fuels—stored organic matter from the distant past. Often these do not contain precious trace minerals essential to many plants and, after a period of high yields, the soil becomes depleted and productivity falls. This in turn requires the application of greater concentrations of fertilizers and more complex mixtures of chemicals.

To obtain higher yields, modern agriculture has to select from highly specialized varieties that depend upon narrowly specified conditions of water supply, fresh doses of fertilizers, and heavy use of pesticides, since they are more vulnerable to attack by pests. In the process the genetic base of the plants itself is made extremely narrow, thereby rendering them more susceptible to pest related diseases. This in turn requires the application of more pesticides—a problem accentuated by the fact that many pests mutate into forms that are resistant to traditional methods of chemical control. New and heavier doses of pesticides are required and they have chemical structures even more alien to living organisms. The run-off of such chemicals is one of the great environmental problems we face today. Also, along with pesticides, modern agriculture utilizes herbicides to destroy all plants, defined as weeds, which are not directly seen to offer agricultural profits. The weeds so destroyed are often precisely those involved in bringing nutrients to the surface. This in turn increases the demand for artificial fertilizers.

At the same time the heavy use of pesticides and herbicides destroys the long-term biological basis of soil productivity. The bacteria, the algae, the worms and other organisms that constitute an integral part of ecological cycles are destroyed or damaged. The soil is no longer able to regenerate itself on its own. It requires help from the outside—mechanical inputs to break up the soil; fertilizers to replenish lost nutrients; new seeds to reproduce genetically selected high yielding crops unable to reproduce themselves; and, above all, fossil fuels for providing the energy, fertilizers, pesticides, and herbicides required.

As a result agricultural systems have been transformed from stable, cyclical, decentralized enterprises into unstable, linear, and interdependent ones. No longer can the farmer assume that, whatever happens elsewhere, the farm itself would be an unaffected self-contained system. It has been integrated dependently into a global network that provides seeds, fertilizers, and pesticides as well as petroleum-energy for agricultural machinery. Such a system is far more vulnerable to global changes than the traditional self-sustaining structure. The traditional farmer depended upon ecological cycles that were local; the new farmer depends on a linear assembly line where fossil fuels have to be continuously imported in the form of energy, fertilizers, pesticides, and herbicides, and agricultural products are sent into centers of population from which the wastes are ejected into the biosphere far away from the soil from which they originated.

Contrast this modern system with traditional agriculture. The latter attempts to bend nature's self-regulating processes rather than depend upon external inputs of energy and fertilizers to maintain soil productivity and fertility; recycles wastes as compost back into the soil so that nutrients are not depleted; and manages pest control by biological methods rather than the use of pesticides or herbicides. By deploying techniques of crop rotation, diversification, poly-cultures, and cover crops not only is the soil better sustained but also the quality and taste of the food grown is better. The process involves using natural technologies to raise soil quality, maintain pest control, retain nutrients, and protect genetic diversity in contrast to the modern system, which creates artificial fertilizers, pesticides, herbicides and new varieties of crops by mechanical techniques of genetic splicing and artificially induced mutations.

Yet, even traditional agriculture involves a separation from the self-regulating powers of nature because it has made human labor replace many activities originally carried out by natural processes. This is precisely why the Daoists were led to critique the changes brought about by the Axial Age agricultural revolution in China. However, if we follow Daoist epistemology and adopt the feminine principle of yielding to nature's way, we would be able to develop a very different model of agriculture. Many cultures, such as the Kayapo in the Amazon Basin of South America, practice what can be described as permaculture. This manages crop production by imitating ecosystems evolving toward a climax, but controlling the climax achieved by guiding carefully the process of succession. Carolyn Merchant describes such a system of agriculture noticed by Holmgren and Morrison:

> In contrast to monocultural agriculture, permaculture uses several stories of trees, shrubs, vines and perennial ground crops to absorb more light and nutrients, increasing the total yield. Plants and animals co-exist in separate niches that reduce competition and promote symbiosis among species. Complexity not only helps to ward off catastrophes but increases the variety of foods produced. External energy and physical labor decrease as perennials mature so that the energy needs are provided within the system. Permaculture is highly adaptable and is applicable to a spectrum of habitats. (Merchant 1992: 214–215)[35]

Permaculture takes one step further the Daoist dictum of following the way of nature. In organic agriculture obeying the dictum involves deploying natural technologies but in conjunction with human technologies to permit intensification of production. In permaculture it involves setting up a system that lets nature proceed with much less human intervention. The only difference is that the system's equilibrium at climax has, because of a judicious selection of crop mixes, been made to be one that may not have developed without human intervention.

Thus, there are clearly three possible approaches to agriculture—the modern system that attempts as far as possible to model agriculture on industrial systems; the tradition of organic farming that combines both human and natural technologies in an integrated fashion; and the approach of permaculture that models agriculture on natural biological systems. The ideal of the Daoists would be permaculture for it would be the ultimate achievement of following the way of nature, even if the final climax is defined by humans through calculated control of the succession processes that lead to it. In fact, it is safe to assume that Laozi's vision of following the way of nature was itself inspired by the permacultural practices of the neolithic cultures that were being displaced by the agricultural revolution in China.

The example of agriculture shows the importance of managing our relations with nature in ways that are sensitive to its inherent self-regulating powers that produce goods and services freely for human beings. The discipline that most directly and crucially shapes our relationship with nature is economic science. Hence, it can be an important area of knowledge that can profit by learning from ecofeminist and Daoist thought, by incorporating into itself epistemological insights inspired by the feminine principle. Current economic theory is rooted in a mechanical vision driven by the masculine principle of dominating nature that leads it to systematically ignore the production of goods and services for humans by self-regulating natural systems. What is needed is clearly an approach that integrates the human production of goods and services in industrial systems with nature's production—an approach that informs the *yang* approach of modern economics with the *yin* approach endorsed by Daoists and ecofeminists.

Yet, there have been few attempts to inform modern economic theory with the feminine principle. Even Gross and Levitt, in their scathing attack on the epistemological realignment of science proposed by ecofeminists, note that although such critiques have "been promulgated with extraordinary success in the humanities and social sciences, even in legal education and research. … Economics has seemed, for some reason, relatively resistant" (Gross and Levitt 1994: 108). The rest of this chapter is concerned with examining the implications for economic science of Daoist and ecofeminist epistemological insights and how these are connected to the epistemology of complementarity.

4.3 Informing Economic Exchange Value with the Yin Principle

It is generally acknowledged that modern economic theory developed within the intellectual horizons defined by classical science and its mechanical vision of nature. Hence it should not surprise us to find that it came to articulate its relations with nature in a fashion that systematically adopted the *yang* orientation of ignoring natural grown properties and self-regulating processes. This also influenced the fundamental exchange laws of economics that were formulated very early in the history of the discipline. These exchange laws were determined without any regard for the role of nature in the production of goods and services for humans. Instead we find that the exchange laws promote an essentially dominative orientation to nature that assigns value to production by humans but not to production by nature.[36]

Take, for example, the law of supply and demand, which determines exchange values in free-market economies, that is, in nearly all of the economies we have today. It completely ignores nature's role in the production of environmental services such as clean air, fresh water, good soils, food in the oceans, a climate suited to human habitation, and so on. Consider, for example, the services provided by a rainforest. These are ignored when the value of the forest is treated as no more than the cost of logging it for human use. The natural environmental capital embodied in the forest, and which is needed to reproduce it so that it can continue its services, is not accorded any recognition. Thus the unrestricted supply-demand law operates without any constraint or commitment to reforestation or sustainable forest management.

The same may be said of the law of labor that had earlier informed socialist economic theory. This theory assumes that the exchange value of a commodity is determined by the amount of labor embodied in it. On such an account, things produced spontaneously by nature have no more value than the amount of human labor expended in appropriating them. The production of goods and services by the natural environment itself is accorded no economic value since no human labor has been embodied in its production.

The laws of exchange value are fundamental to economic science, and the discipline itself could be said to begin with these laws. The reason they have been formulated and applied with such utter disregard for natural environmental capital is that they were originally developed in the historical context of the industrial revolution, and under the inspiration of a mechanical conception of nature and production in which the distinctive structure of grown properties are not recognized. Prior to the rise of the industrial market economy most human beings lived essentially as agricultural farmers, pastoralists, or hunter-gatherers. In many societies trade played a role of course, but it essentially involved luxury goods intended for ruling elites; the majority of the population mostly consumed what they themselves produced in the local regions where they lived and worked.

The rise of the modern market economy transformed this system. Two key factors shaped the genesis of modern economic theory—first, the view of production as industrial manufacture in a factory deploying mechanical technologies; and second, the role of the market as a mechanism for distributing the products of industry.[37] Producers no longer consumed only what they produced; consumers no longer produced most of the goods required to meet their basic needs. Production and consumption became sundered as the former was geared toward making products for sale; and the basic needs of people were met by buying what was needed at the marketplace. The market itself became the mechanism for distributing the goods produced to consumers who wanted them.

Two essential problems confronted the economic philosophers and thinkers who studied the emerging market system. The first was to describe the natural exchange laws that operated in the marketplace; the second was to make the marketplace conform to the natural laws of exchange value thus discovered. Indeed the descriptive and prescriptive problems were intertwined—the descriptive laws themselves were intended to mold, shape, and direct market activity.

The reason for conflating the descriptive and prescriptive problems can be understood once we realize that the presence of a market by itself does not uniquely dictate the framework of exchange values under which it will operate. Commercial markets have existed—and large ones at that—in very different areas of the world in various historical periods. Prior to the modern age a massive market system linked diverse cultural zones from Europe, North and East Africa, the Middle East, South and Southeast Asia, and East Asia.[38] But in contrast to the modern market system these market transactions were subject to the specific religious, cultural, and sociopolitical goals of the various cultures they connected. The idea of a market that operated as an autonomous system had yet to evolve. The idea of human beings as producers and consumers only—this economic reductionism of human beings in the ontological and commercial sense—is the distinctive invention of the modern era.

This notion of humans as producers and consumers—albeit the same people in different roles—meeting at the marketplace to exchange the fruits of their efforts lies at the heart of modern economic theory. From the beginning of the discipline in the modern era it raised the question of what should be the fairest way of managing the exchange—that is, of deciding on the exchange value of the products in the marketplace. The answers that came to be given were to have a crucial impact not only on subsequent economic thought, but also on the social and political history of the societies that emerged after the industrial revolution.

Central to the solution of the problem of determining the exchange values of commodities and goods produced for the marketplace were the egalitarian ideas that emerged in the Enlightenment—the notion that all human beings were to be treated as having equal political and economic rights. These egalitarian notions were not questioned by economic theorists but taken for granted by them. They were presuppositions that informed their approaches to the question of defining exchange value. However, there is an ambiguity in treating human beings as equal when they are seen as economic agents? In what respect are they equal? Are they equal because they have equal rights as consumers of economic goods and services? Or, are they equal because they have equal rights as producers of such goods and services? This dilemma was never resolved by Enlightenment philosophers to the satisfaction of everyone. Instead two conceptions of exchange value emerged—one based on consumer egalitarianism which led to the law of supply and demand as the basis for establishing economic exchange value, and the other based upon producer egalitarianism dependent on the law of labor as the determinant of exchange value. Each came to have a profound influence on the future social and economic development of humankind.

Let us examine the solution that emerged from the assumption of consumer egalitarianism—the law of supply and demand. This law was derived by asking "What is the fairest way of distributing the goods and services provided by the market to a body of equal consumers?" The answer was to treat the market as a place in which commodities were auctioned and all persons were allowed to bid freely the price they were prepared to pay. Then goods in short supply would be expected to fetch higher prices; so would goods in high demand. Conversely, if demand were to fall, or supply rise, prices would correspondingly come down. The price at market equilibrium would embody consumer justice, that is, it would define the just price consumers should pay when treated as equals.

On the other hand, the exchange value could be decided by approaching the problem from the point of view that human beings are equal producers. If many producers meet at a marketplace to exchange the products of their labor, then the fairest way of allocating the rewards for production would be to set the exchange value for a product in terms of the amount of labor involved in producing it. A price exceeding the labor involved in its production would be unfair to others; a price below the amount of labor involved would be unfair to the producer. Only the law "to each according to his labor" would reflect producer justice.

Of course classical economic theorists did not perceive themselves as prescribing how prices ought to be established in the market, but saw themselves as merely describing how they are established through the processes of market equilibration. They perceived themselves as discovering, rather than constructing, the exchange law that operated within the marketplace in the same way that natural scientists, upon whom they modeled themselves, discovered natural laws in the physical realm.[39]

But in thus perceiving their accomplishment they only deluded themselves. They masked their role, or the role of their egalitarian conception of consumers and producers involved in market exchanges, in determining the exchange laws that were made to condition economic activity. After all market activity *per se* can occur under many different conditions with different regulating constraints. Prior to the modern era it was constrained by social, cultural, religious, and even ecological requirements. The philosophers and social critics of the Enlightenment had to wage a ferocious battle to remove these constraints. And a great part of the battle was won when the market was described solely as driven by a law of exchange value—whether the law of supply and demand or the law of labor—in which all properties were abstracted from human beings within a marketplace, except that which defined them as either equal consumers or equal producers.

Clearly the laws of economic exchange value have not been derived in quite the same way as natural laws in science. The moral basis of their origin in egalitarian ideals suggests that they are rather expectations to which economic activities are required to conform. Of course, these expectations may themselves have been abstracted from a particular market system—that is, a market that treated all consumers, or producers, as equal—but what is important is that they could subsequently be made to regulate market activity everywhere to increasingly conform to their expectations.

It is precisely because they are expectations that economists can talk of failures of the market mechanism when prices do not conform to the law of supply and demand; or unfairness when they do not reflect the law of labor. It would be very strange if physicists were to similarly speak of failures in the law of gravity or the law of conservation of energy, and require that natural systems be adjusted to conform to these laws. Natural laws, unlike the laws of exchange value, are not expectations and can never, if they are true, be violated.[40] This does not, of course, mean that the exchange laws do not describe, but they do this precisely because the system itself is also largely regulated by these laws. The laws function as expectations both with regard to how the world actually behaves, and how it should behave.

Treating economic exchange laws as expectations we have imposed on the way exchange values are established raises the question of the extent to which these laws value the production and distribution of goods and services by human beings over that of nature. Do these laws of exchange value acknowledge the yielding feminine principle of not violating the self-regulating natural systems that also produce goods and services for humans? The categorical answer is that they do not. Indeed the exchange laws are indifferent to natural contexts because they cannot acknowledge nature's production of goods and services. They reflect a one-sided orientation of only paying regard to the assertive masculine principle of domination over nature that recognizes only the human production and human distribution of goods and services.

Consequently the law of labor only sees economic value in the human effort expended in production, and the law of supply and demand only considers the human role in the supply of, and the demand for, a product, without regard to its ecological impact. Both laws are indifferent, and often antagonistic, to sustaining natural environments and their self-regulating potential. Hence, the unconstrained operation of these laws of exchange value precludes acknowledging and respecting the natural contexts that manage production and distribution of indispensable goods and services that make human life and well-being possible. It valorizes the masculine principle without giving any role for the feminine principle in defining our relations with nature.

In order to appreciate this point more fully let us return to the origins of modern economic theory by re-examining the views of Adam Smith on exchange value. It might appear that economic theory has advanced so much since his time that his views have no more central relevance. But this is to miss the key point—the problem is not in the refinements of economic theory but the fundamental first principles that established the discipline, and upon which later economists have erected their theoretical edifices. In this regard Smith is of paramount significance for a number of reasons. First, with the demise of socialism his views have achieved unparalleled influence, directly or indirectly, on current economic thought. Second, he was one of the earliest economists to attempt to lay a rigorous foundation for the discipline and he wrote at the dawn of the industrial era. Third, his account of the division of labor has had a great impact on the articulation of industrial structures in modern societies. Fourth, he was one of the earliest thinkers to make a sharp distinction between the 'use value' and 'exchange value', or price, of a commodity and to give an account of how exchange value gets established. Finally, and most intriguingly, in his epochal study *The Wealth of Nations*, we see two attempts by Smith to provide an account of how exchange value arises.[41] One leads to the socialist labor theory of value, and the other to the free market law of supply and demand. Since the subsequent development of economic thought adopted one or the other of these two notions of exchange value, it is instructive to examine the arguments Smith presents to support each of these positions. We find that, in both cases, the role of nature in production is ignored and only human action is recognized as productive—an attitude criticized by feminists who indict modern science for its project of adopting a patriarchal attitude of dominating and degrading nature, and Daoists who see it as ignoring the Way of Nature. It also disregards what Bohr referred to as anthropological complementarity by giving weight to the role of culture (humans) in the process of production, but not the complementary aspect of nature.

Smith begins by arguing that the original law of value that guided economic exchanges in human cultures was the law of labor. Assuming the division of labor, and the necessity of exchange in any society that develops beyond the most primitive, Smith derives the labor theory of exchange value as follows:

> Every man is rich or poor according to the degree in which he can afford to enjoy the necessaries, conveniences, and amusements of human life. But after the division of labor has once thoroughly taken place, it is but a very small part of these with which a man's own labor can supply him. The far greater part of them he must derive from the labor of other people, and he must be rich or poor according to the quantity of that labor which he can

> command, or which he can afford to purchase. The value of any commodity, therefore, to the person who possesses it and who means not to use or consume it himself, but to exchange it for other commodities, is equal to the quantity of labor which it enables him to purchase or command. Labor, therefore, is the real measure of the exchangeable value of all commodities. (Smith 1961: 34)

Of course there are complex issues associated with determining the amount of labor embodied in a product—we have to take into account the time expended in making the product, the labor involved in acquiring the skills needed for its production, the intensity of labor employed and so on. Objective criteria for measuring these are not easy to establish and they lead to numerous problems in founding a quantitative economic theory upon the law of labor. Smith himself was motivated, in part, to go beyond the labor theory because of this problem of the heterogeneity of labor. Yet, Marx was to return to the labor theory and produce a highly influential critique of capitalist society.[42] Moreover, similar problems arise in the context of utilitarian theory and welfare theory—problems related to quantifying the basic notions of utility and welfare. Hence, even if the quantification problem is difficult, it is possible to set it aside as something all economic theoreticians face, and discuss the arguments for and against adopting a particular notion of exchange value.

But the claim that the value of a commodity to a seller is equal to the amount of labor it enables him to purchase does not establish that it is always the case that the labor involved in its production is equal to the labor embodied in the commodity it is exchanged for. For example, a thirsty person in desperate need with no private access to water would be prepared to exchange a piece of furniture he has made at great expense of labor for a glass of water. Clearly all exchanges need not conform to the law of labor. In this respect also the laws of exchange value are different from the laws of physics, which cannot be violated under any circumstance whatsoever.

The law of labor follows only if the exchange is between participants who in certain respects are equal, that is, each is able to produce either of the commodities being exchanged but, because of the efficiency inherent in the division of labor, they decide to specialize so that each produces only one kind of commodity and meets the need for the other by exchange. In that case they are likely to decide that only embodied labor can reflect the appropriate exchange value.

Smith illustrates how the labor theory of exchange value could arise in his genesis myth of "The hunters, the deer and the beavers":

> In that early and rude state of society which precedes both the accumulation of stock and the appropriation of land, the proportion between the quantities of labor necessary for acquiring different objects seems to be the only circumstance which can afford any rule for exchanging them for one another. If among a nation of hunters, for example, it usually costs twice the labor to kill a beaver which it does to kill a deer, one beaver *should* naturally exchange for, or be worth, two deer. It is natural that what is usually the produce of two days' or two hours' labor *should* be worth the double of what is usually the produce of one day's or one hour's labour (Smith 1961: 53). [My emphasis]

It is important to note that Smith is not at all concerned with whether the value of a deer should be a great deal more than a beaver if it takes nature a greater amount of time to reproduce a deer. Nature's production of deer is assigned no value; only the human labor expended in hunting and killing it is given economic value. Such a notion of exchange value would allow a hunter to exchange the last beaver in the world for two deer, say—the price being determined solely by the labor involved in killing them. This absence of constraint of any sort, which accords some recognition to the reproductive powers of nature, is at the heart of what informs the labor theory of value. It contrasts sharply with the feminist call to respect natural contexts and its reproductive powers.

There is a normative judgment in Smith's derivation of the labor theory that he fails to recognize. He seems to suppose that he derives his law of labor as one that is natural, in the sense of being obedient to natural law. However, it is a most peculiar sort of natural law, one that can be violated but which *should* be (the prescriptive mode of presentation is Smith's, see quote above) conformed to. This normative judgment, cast in the guise of a description, leads the economic theoretician Joan Robinson to conclude that Smith is not involved with mere description, but with establishing a natural price in the ethical sense of the medieval notion of 'just price' (Robinson 1973: 30–31).

Smith introduces two caveats to his labor theory. These do not detract from the theory but add to it by making it conform more directly to our intuitions. The first is that in determining the amount of labor embodied in a product we need to take into account the severity of the work involved. Allowance must be made for the intensity of labor deployed so that an hour of labor of intensity one unit is equal to two hours of labor half as intense. Second, allowance should be given to labor requiring great dexterity or ingenuity since all such skills are the result of prior effort made in learning and training that have been transferred to the present context. Smith therefore concludes:

> In this state of things, the whole produce of labour belongs to the labourer; and the quantity of labour commonly employed in acquiring or producing any commodity, is the only circumstance which can regulate the quantity of labour which it ought commonly to purchase, command, or exchange for. (Smith 1961: 54)

Smith, however, proceeds to argue that the labor theory of value ceases to be applicable beyond what he calls the primitive stages of society. With the rise of private property, or the accumulation of private capital, the law of labor breaks down and new factors come into play affecting the exchange value. The accumulation of private capital allows the owner to rent his property or to purchase the labor of others, rather than the produce of their labor. Labor itself becomes a commodity like the products of labor. This commodification of labor, often treated as a discovery of socialist thought,[43] is already recognized by Smith, albeit in a germinal form since he does not develop the subsequent notion of labor power so central to the economic theory of Marx.

In order to demonstrate the limits of the labor theory of exchange value, once labor itself is made into a commodity, Smith gives the following argument. He considers what happens when owners of capital employ workers and offer them wages for their labor. The employer also works, but this is the effort involved in the inspection and direction of the work of production. Smith then shows that the profits, or rewards for the labor of the employer, is determined more by the stock of capital he employs than the labor he puts into his work. Thus, two manufacturers may each employ 20 workmen at 15 pounds per year. If the cost of material and infrastructure for the manufacturer in one case is 700 pounds, and in the other case is 7000 pounds, then their total cost would be 1000 pounds and 7300 pounds, respectively. This takes into account the annual wage bill of 300 pounds for each manufacturer that is paid to his workers. However, their profits at a return of 10 % per annum would be 100 pounds and 730 pounds respectively. Thus, the employers' returns do not correspond to the labor they put in to carry out their tasks of inspection and direction, but to the amount of capital they invest in their enterprise.

Similarly, the rise of private property also allows owners to rent out their land to those who wish to harvest its natural produce. This also causes prices to deviate from the law of labor. In fact, it adds a third factor of production—rent—beyond capital and wages to determine prices or exchange values. Thus, the natural produce of the land such as "the wood of the forest, the grass of the field and all the natural fruits of the earth," which originally cost the laborer only the effort of collecting them, now have an additional charge attached to them. The rent, being the price of the license for gathering the natural produce, adds a cost over and above the labor involved in harvesting them.

These changes wrought by the rise of private property, lead Smith to conclude that the notion of labor as a measure of exchange value cannot apply beyond the primitive stages of society. Exchange value has now to be effected by a different route—one that leads him to the law of supply and demand:

> The quantity of every commodity brought to market naturally suits itself to the effectual demand. It is the interest of all those who employ their land, labor, or stock, in bringing any commodity to market, that the quantity never should exceed the effectual demand; and it is the interest of all other people that it never should never fall short of that demand.
>
> If at any time it exceeds the effectual demand, some of the components parts of its price must be paid below their natural rate. If it is rent, the interest of the landlords will immediately prompt them to withdraw a part of their land, and if it is wages or profit, the interest of the labourers in one case and of their employers in the other, will prompt them to withdraw a part of their labour or stock from this employment. The quantity brought to market will soon be no more than sufficient to supply the effectual demand. All the different parts of its price will rise to their natural rate, and the whole price to its natural price.
>
> If, on the contrary, the quantity brought to market should at any time fall short of the effectual demand, some of the component parts of its price must rise above their natural rate. If it is rent, the interest of all other landlords will naturally prompt them to prepare more land for the raising of this commodity; if it is wages or profit, the interest of all other laborers and dealers will soon prompt them to employ more labor and stock in preparing and bringing it to market. The quantity brought thither will soon be sufficient to supply the effectual demand. All the different parts of its price will soon sink to their natural rate, and the whole price to its natural price.

> The natural price, therefore, is, as it were, the central price, to which the prices of all commodities are continually gravitating. … The whole quantity of industry annually employed in order to bring any commodity to market, naturally suits itself in this manner to the effectual demand. It naturally aims at bringing always that precise quantity thither which may be sufficient to supply, and no more than supply, that demand. (Smith 1961: 65)

It is evident that Smith who began with the notion of a hunter community of egalitarian producers, and thereby made the labor involved in production the ultimate determinant of price or exchange value has, as a consequence of the existence of private property, moved toward a law of supply and demand as the effective determinant of price. In the process, he raises numerous issues that were to be taken up by subsequent economic theoreticians. The socialists enlisted the labor theory of value as the natural measure of economic exchange and called for the abolition of private property—the root cause identified by Smith as leading beyond this natural law. Others defended private property and free markets by making the law of supply and demand a natural outcome of both. Indeed the fundamental conflicts that subsequently arose within industrial society after Smith can be traced to these two divergent ways of establishing the law of exchange value for economic practice.

But it is also noteworthy that in shifting from the labor theory of exchange value to that of supply and demand, Smith has not relinquished his fundamental indifference to nature's role in the production of goods and services, and the reproduction of the natural capital that makes this possible. He does not at all take into account nature's role in production or its capacity to reproduce the products that are extracted from it. The law of supply and demand is merely concerned with the supply of a commodity relative to the human demand for it—if there is effective demand for deer meat then nothing in the law restricts meeting it even if it involves killing the last deer able to reproduce the species.

In formulating the law of supply and demand, Smith has merely changed his perspective from that of equal human producers to equal human consumers. He has adopted the notion of a free market, or competitive model of equal consumers, who determine prices by their bids. If we assume a model of consumers all of whom have perfect and free information of market transactions, and each of whom have only a minimal influence upon the market—the so-called conditions that Charles Dyke describes as required for "consumer sovereignty" (Dyke 1981: 134)—then prices would gravitate, as Smith has argued, to the natural prices determined by supply and demand. What is important is that everyone enters the market as equals, and is not discriminated against. Clearly, this assumption embodies a conception of justice—or just price—just as the law of labor. Dyke emphasizes the point:

> Any discrimination between market participants willing, say, to make offers for a product is thought to be an imperfection in the market ... Our sense of justice rebels against this. We insist that everyone be treated equally, hence, anonymously and as an abstraction: "potential buyer". (Dyke 1981: 139)

However, this consumer ethos—humans as merely "potential buyers"—is precisely what has to be questioned if nature is a self-regulating system with a limited capacity for reproducing itself. Mere considerations of available supply and human demand cannot be made the sole criteria for determination of prices, or even whether a commodity should enter the market. Ecological considerations must come into play that would place restraints on the free play of the supply-demand law, and the operation of market forces, if we are to preserve the self-regulating powers of nature. Such ecological restraints would incorporate the feminine principle of yielding to nature's contexts to moderate the masculinist formulation of economic exchange values.

Smith's unrestrained market model would apply only in the case of products that are assembled in a factory and enter the market; that is, they are products made by humans deploying mechanical systems. Then it is possible to suppose that we can raise production to provide the requisite supply whenever demand increases. In the case of natural products, say timber and fresh water, such an assumption is tenable only if we assume that their supply is infinite or unlimited, or that it is always possible to create new technologies to meet rising demand.

However, these assumptions implicitly made by Smith, who does not even seem to recognize a problem here, are highly questionable. While they may apply to manufacture they cannot be equally extended to ecologically produced goods and services. Production by nature cannot be modeled on manufacture by machinery. Self-regulating processes make natural products. The ecosystems involved have a finite and limited capacity for reproduction. Moreover, once degraded or destroyed, natural systems cannot be restored by human beings however great the demand becomes, for the knowledge to do this is not human knowledge but information embodied in nature. The dodo bird cannot be reproduced regardless of demand since we have lost forever the information embodied in the bird's genes and its habitat.[44]

There was a wider recognition of the limits that needed to be imposed on human extraction of natural produce in premodern societies more in touch with organic processes in nature. Such societies were able to see that human extraction of natural products must give regard to nature's role in their production. According to De Klemm, many traditional societies imposed limits in the interests of conservation:

> [In] most traditional systems … life support systems were preserved because sustainable forms of land use such as terracing, stable shifting cultivation and moderate pastoralism were usually practiced. The harvesting of wild animals and plants was governed by religious beliefs and customary rules that made it sustainable. Genetic diversity was maintained as a result of low pressure exercise over natural systems and by the imposition of religious taboos or the existence of sacred groves or ponds. (De Klemm 1985: 245–246)

But it was precisely these limits that were historically lifted when agricultural society was transformed into the industrial order as a result of the socioeconomic revolution, both swift and dramatic, mediated by giving free rein to the law of supply and demand. In the process the feminine principle of yielding to nature that also restrained economic activity came to be relinquished, and economics came to be swayed completely by the masculine principle of dominating nature. In his study of the process of the emerging free market system Karl Polanyi describes this change:

> If from the outset the logically fallacious identification of "economic phenomena" and "market phenomena" was understandable, it later became almost a practical requirement with the new society and its way of life which emerged from the industrial revolution. The supply-demand price mechanism whose first appearance produced the prophetic concept of "economic law", grew swiftly into one of the most powerful forces to enter the human scene. Within a generation—say, 1815 to 1845, Harriet Martineau's "Thirty Years Peace"—the price making market, which previously existed only in samples in various ports of trade and stock exchanges, showed its staggering capacity for organizing human beings as if they were mere chunks of raw material and combining them, together with the surface of mother earth which could now be freely marketed, into industrial units under the command of private persons mainly engaged in buying and selling for profit.

> Within an extremely brief period, the commodity fiction, as applied to labor and land, transformed the very substance of human society. Here was identification of economy and market in practice … The true scope of such a step can be gauged if we remember that labor is only another name for man, and land for nature. The commodity fiction handed over the fate of man and nature to the play of an automaton that ran in its own grooves, and was governed by its own laws. (Karl Polanyi 1977: 9–11)[45]

Polanyi encapsulates succinctly the process that led to the almost overnight emergence of capitalist industrial society and the problems and issues it subsequently raised into prominence. Shortly thereafter the Marxist critique emerged to concentrate on the commodification of labor. It returned to the law of labor, and projected the evolution of the free market economy to a point where the increasing concentration of wealth in a capitalist minority, and the systematic and degrading pauperization of the laboring class, would result in the revolutionary overthrow of the capitalists. As a result there would be a change in the ownership of the means of production from private individuals to society as a whole. By eliminating private property, the socialists argued, we can once again reward each according to his or her labor, as was the practice in primitive cultures. What Adam Smith saw as the failure of the law of labor beyond primitive society, because of the rise of private property, would be resolved not by moving beyond to a new law of exchange value, but by doing away with private ownership altogether.

However, by failing to address what Polanyi calls the "commodity fiction" applied to land, that is nature, socialist thought also promoted another variant of industrial society in which nature's role in production came to be ignored. Resolving this problem requires going beyond the question of the private or public ownership of the means of production that defined capitalist versus socialist debates—it requires raising questions concerning the nature of the means of production.[46] Ecofeminists and Daoists address this question by turning to the feminine principle in order to respect the integrity of natural contexts that provide goods and services needed by humans. They show us that mechanical production in a factory is only one means of producing goods and services—an approach defined by culture—and that there is another means of production connected with nature dependent on its self-regulating technologies. This suggests that economic science has to face the question of how the two technologies may be integrated together. It requires the discipline to adopt the feminine principle of yielding to nature, and placing limits on the unrestrained application of the economic laws of exchange value, so as to sustain self-regulating processes in nature. It requires it to see the feminine (*yin*) approach of yielding to natural contexts that shape growing things as complementing the masculine (*yang*) approach of shaping natural contexts to serve human ends that has inspired economic theory in the modern era. In the next section we examine how such an integrative approach can be implemented within the framework of the exchange laws of economic science by informing it with ecofeminist and Daoist insights, and the way Bohr's anthropological complementarity reflects this synthesis.

4.4 Natural and Human Capital: Anthropological Complementarity and Economic Theory

We have seen Bohr emphasizing the significance of explaining human behavior in terms of the anthropological complementarity of nature and nurture—one that can also be seen as acknowledging the feminine principle of yielding to nature and the masculine principle of controlling nature. This is possible because there are not only goods and services sustained by economic activities where humans assert their actions over nature but also others produced by nature, which are guided by ecological self-regulating processes.[47] The former may be seen as production by humans, and the latter as production by nature. One of the earliest writers to criticize economic science for ignoring the role of nature in production was Schumacher. He developed his critique in his influential book *Small is Beautiful*[48]:

> Modern man does not experience himself as a part of nature but as an outside force destined to dominate and conquer it. … One reason for overlooking this vital fact is that we are estranged from reality and inclined to treat as valueless everything that we have not made ourselves. Even the great Dr Marx fell into this devastating error when he formulated the so-called 'labor theory of value'. Now we have indeed labored to make some of the capital which today helps us to produce—a large fund of scientific, technological and other knowledge; elaborate physical infrastructure, innumerable types of sophisticated capital equipment etc.—but all this is but a small part of the total capital we are using. Far larger is the capital provided by nature and not by man—and we do not recognize it as such. This larger part is now being used up at an alarming rate, and this is why it is an absurd and suicidal error

> to believe, and act on the belief, that the problem of production has been solved. (Schumacher 1974: 10–11)

Schumacher is making a distinction, already recognized by Laozi, between production managed by humans and production managed by nature; between "the Dao that can be told of" and "the eternal Dao." However, living in the industrial age when nature itself is under threat he no longer speaks of nature's way as being eternal. Moreover, Schumacher introduces the duality of human and natural production in a language alien to Laozi's philosophy. He speaks in the economic vocabulary of capital—the stock of capital created by humans (property, physical equipment, infrastructure, knowledge, etc.) and that provided by nature. He argues that though we value human capital—capital nurtured by us—we seem to assign no value to natural capital.

Let us now consider Schumacher's approach in the context of the application of the law of supply and demand because the law of labor is hardly used today to establish exchange values. We have seen that the law of supply and demand was derived by considering the fairest method of distributing goods and commodities to equal consumers who meet at a marketplace. However, the original derivation assumed that we need to be concerned only with the synchronic dimension of fairness. Once we realize that goods and commodities are also produced by natural capital, we also have to acknowledge the possibility that one generation of humans could unfairly consume natural capital at the expense of succeeding generations. They could do this by both overexploiting and polluting the natural world. As a result the question of fairness also requires taking into account the diachronic dimension of application of the law by elevating to the center the problem of intergenerational justice.

The environmental economist J. A. Butlin raises this issue:

> Encapsulated in environmental and natural resource management is the problem of time. It permeates each resource utilization problem with which man is faced. Neoclassical capital theory incorporates the time dimension in a quite adequate way, from the point of view of the current generation … but can we trust myopic, efficiency based, self-interest maximizing criteria to allocate resource use between generations, particularly if the allocation is from a fixed stock of exhaustible resources? (Butlin 1981: 60)

It is reasonable to suppose that intergenerational fairness requires us to leave future generations at least as much of nature's self-regulating capital as we have inherited. This means we must limit the rate of consumption of natural capital to a sustainable value—that is, it cannot be exploited as a source of raw material, or a sink for absorbing our waste products, beyond its capacity for self-renewal.[49] For example, rivers cannot be exploited for water, or as sinks for factory pollutants, at a rate that degrades them permanently for later generations.

Given the significance of such a diachronic perspective one may well ask why classical economic philosophers failed to consider this wider perspective of justice when they articulated their notions of exchange value. Why did the obvious question of intergenerational equity not enter into their calculations despite their deep concern for equity of consumers and producers in the determination of exchange values? One reason for this oversight can be attributed to the fact that their studies were a response to the vast expansion of manufacturing activities that ushered in the industrial revolution. Consequently, their notions of production were shaped to a large extent by their study of manufacturing activities deploying industrial techniques in which self-regulating processes were never encountered.

Their view of manufacture as the archetypal mode of production precluded classical economists from perceiving the distinctively different mode of production in which natural self-regulating systems played a crucial role. As a result they also failed to recognize that natural technologies, and the capital they carry, could become irretrievably lost to future generations without conscious effort to preserve and sustain them. We have seen that natural technologies operate on principles generally opaque to the people who benefit from them; that they are difficult, often impossible, to replace once lost; and that substitutes for them are often not available. Hence natural technologies cannot be treated in the same way we treat mechanical or industrial technologies—the latter, unlike the former, can be replaced even if destroyed because human knowledge, enterprise, and effort would be sufficient to accomplish this task.

Hence, the unrestricted application of the law of supply and demand, modeled on our dealings with industrial systems, cannot be deployed in an unrestrained fashion to manage natural technologies. We need, in addition, ecological criteria that would restrain economic exploitation within limits so that natural technologies would be sustained for future generations. Such criteria would have to preserve natural production systems precisely because *we know that we do not know* how natural environmental capital reproduces itself; though we know that left to itself it is able to do so. This is to recognize both the Daoist principles of the knowledge of non-knowledge and the action of non-action. It is to acknowledge that we need to preserve and sustain natural modes of production for future generations because, once lost, they cannot be reproduced by us, simply because we have no knowledge of how to do it.

Indeed, an industrial model of technology inspires many of those who argue for complacency with regard to sustaining natural systems. It leads them to believe that, by allocating sufficient economic resources, we would be able, should the need arise (i.e. with sufficient demand), to nurture back to health, or even nurture anew, degraded or destroyed natural systems—or find new ways of achieving what they now accomplish for us. This is to presume that economic demand and technological advance, the second being a response to the first, would always guarantee future generations a solution to any problems we might create now by our destruction of natural environmental capital.[50]

Such indifference is dangerously misplaced given the vast difference between natural and manufacturing technologies. Manufacturing technologies can always be reassembled because we have detailed knowledge for performing such a task. Moreover, we have reason to be optimistic about achieving progress with such techniques because there is a history of advances in this area—these technologies have increased in diversity, complexity, and sophistication over time. But the same cannot be said of natural technologies. In spite of all the breeding experiments in farming that have taken place over the last four hundred years we have not managed to produce a single new species—and this is not for want of trying.[51] History hardly gives us reason to be optimistic about the prospect of re-creating, or finding substitutes for, natural systems that we have degraded beyond repair or destroyed—quite the contrary. Arguments for complacency, which appeal to future advances in technology, lack what Laozi knew—knowledge of our non-knowledge.

Hence, fairness to consumers in future generations demands that we must adopt the *yin* approach of sustaining the natural capital and natural technologies we have inherited from past generations. Moreover, we cannot adopt the standard economic practice of discounting natural capital in the future the way we discount manufacturing capital we have nurtured ourselves so as to take into account both its wear and tear, and also new technological advances. Since natural capital is self-reproducing, provided it is managed in a sustainable way, it does not suffer wear and tear; being a system whose principles of operation are largely opaque to us we cannot expect technological advances to replace lost natural systems.

Moreover, adopting the *yin* approach of imposing ecological constraints on its exploitation affects the exchange value we assign natural capital. Thus ecologically defined constraints on the exploitation of natural systems can be used in conjunction with standard *yang*-inspired economic theories to sustain such systems. Consider a forest. When we ignore its role as capital in reproducing itself, its exchange value is no more than the cost of the land and the cost of reducing it to raw material—for example, its value as timber—because no human labor or knowledge was involved in creating the forest. Yet in moderating global temperatures it could be performing a vital function.[52] However, our exchange law of supply and demand is not sensitive to this because it can only recognize production as significant if it is accomplished by human beings—it is blind to what the forest produces for us through natural technologies. However we can make the exchange law sensitive to natural production by committing ourselves to sustaining a natural resource base—the criteria for sustainable exploitation being defined by ecological science. Once such criteria are established natural resources immediately acquire an enhanced economic exchange value. The cost of removing timber from a forest reserve that has to be sustained would no longer be merely the cost of its extraction—it would also include the cost of reproducing what has been extracted.[53] This would ensure that the forest resources are exploited within sustainable limits.[54]

Thus, a rational strategy of management of ecological resources requires us to first make a *yin*-inspired commitment to their sustainable use before undertaking a *yang*-oriented economic evaluation. Consumer fairness both across generations and within a generation—what we may call intergenerational and intragenerational justice—can only be established by recognizing the role of both natural and human production, nature and nurture, or natural and economic capital.

Perhaps the most sustained articulation of how to accomplish this task is made in the study *Natural Capitalism: Creating the Next Industrial Revolution* co-authored by Paul Hawken, Amory Lovins and Hunter Lovins. They attempt to articulate more systematically Schumacher's notion of natural capital. They argue, in particular, that future economic development will be crucially dependent upon both the availability and effective functioning of natural capital—especially its life-supporting services that have yet to be assigned any market value. This can be rectified only by designing better business systems, controlling unchecked population growth, and eliminating wasteful consumption patterns. Such a change would require us to include natural capital along with human, manufactured, and financial capital, and improve resource productivity and ameliorate global inequities. Ultimately, they argue, human welfare is best served not by merely increasing total dollar flow, but by improving both quality and flow of services from natural capital.[55]

By following these recommendations we will combine *yin*-inspired ecological criteria to sustain natural environmental capital and *yang*-inspired economic criteria to deal with economic capital (the capital we have nurtured through our technologies). Such an orientation can also be said to learn from Daoist epistemology by respecting both "the eternal Dao" and "the Dao that can be told of". Moreover, in approaching natural technologies through ecologically motivated sustainability criteria, rather than economic criteria founded upon economic exchange values alone, we would be paying regard to the fact that natural systems embody information and principles of which we have no knowledge—except the knowledge that these principles regulate nature without our intervention or knowledge. We would, in effect, be acknowledging what the Daoist epistemological paradoxes teach us—the paradoxes of the knowledge of non-knowledge, the morality of non-morality, and the action of non-action. Thus Laozi's *yin*-inspired epistemology can be a significant corrective to contemporary *yang*-inspired economic science—a discipline motivated by a mechanical vision of nature and industrial modes of production that has yet to adequately recognize the distinctive structure of self-regulating processes and grown properties in nature.

Clearly this also can be seen as taking into account the complementarity of the two approaches of nurture and nature Bohr saw as relevant in the anthropological context of understanding human nature, and extending it to the context of human interactions with nature. It leads us toward a new ecological economics that sees the natural processes in ecology and the nurtured processes in economic production as mutually exclusive, but complementary, ways of producing goods and services essential to human survival. It is even symbolically reflected in the motto "Opposites are Complementary" Bohr selected to go with the female-male *yin-yang* logo for his crest when he received his nation's highest honor.

Notes

1. The concept of *yin* and *yang* in Chinese thought is used to refer to complementary opposites that exist within a larger dynamic whole. It is generally symbolized by the *Taijitu* symbol which expresses the fact that the complementary *yin* and *yang* aspects of a thing may exist as opposites but are not opposing—light and darkness are opposites but not opposing

for one cannot exist without the other. The notion of polar opposites or contrary forces which are nevertheless interconnected and interdependent pervades many aspects of classical Chinese science, philosophy, medicine, and the martial arts.

2. See Shankman and Durrant (2002), Lloyd and Sivin (2002). Also Bala (2006), pp. 127–128.
3. According to Wing-Tsit Chan (1963) a system called *ko-i* was developed to match concepts between Buddhism and Daoism. This not only facilitated the understanding of Buddhist concepts but also, in certain respects, transformed Buddhism so that it became "essentially Chinese in both thought and language" pp. 336–337.
4. The Chinese word Tao or Dao means "way," "path," or "route," but it is sometimes deployed to refer to "principle" or "doctrine." As a metaphysical concept it was first introduced by Laozi, becoming a central notion for both religious and philosophical Daoism. The concept of Dao subsequently came to be adopted by both Confucian and Zen Buddhist thinkers.
5. The *Dao De Jing* has been translated over 250 times into various European languages, especially English, German, and French. See Michael LaFargue, and Julian Pas, "On Translating the Tao-te-ching," in Kohn and LaFargue (1998), p. 277. Even in Chinese there are a number of transmitted editions in historical times, with the three primary ones named after early commentaries of the text—the "Yan Zun Version," attributed to the Han Dynasty scholar, Yan Zun (80 BCE-10 CE); the "Heshang Gong Version" named after Heshang Gong (202-157 BCE); and the "Wang Bi Version" named after Wang Bi (226–249 CE).

   However, recent archeological discoveries have unearthed manuscripts some of which antedate the historically received texts. These include the Mawangdui Silk Texts, dating from 230 to 210 BCE discovered in a tomb in 1973. Even older are the Guodian Chu Slips written on bamboo tablets, found in a tomb close to the town of Guodian in 1993. However both these recent discoveries are not inconsistent with the received texts except for chapter orderings and character variance.
6. Quoted in Needham (1956) p. 33.
7. According to Sinologist and philosopher, H. G. Creel (1953, p. 113) the Confucians saw an ordered system of government as one controlling both physical and human nature for the benefit of society as a whole. He also notes:

   > In the midst of our cities Daoism may well seem nonsense. But go out to nature, the trees, the birds, the distant view, the placidity of a summer landscape or the savage fury of a storm, and much of Daoism will seem to possess a validity stronger than that of the most intricate logic. (p. 101)

8. In contrast to contemporary environmentalists who are responding to the threat posed to natural ecosystems by pollution and overexploitation, Daoists saw the threat not as directed at nature but at humans.
9. See the Chan (1963) translation of the *Dao De Jing*. pp. 139–176.
10. *Dao De Jing*. Trans. Chan (1963) p. 175.
11. In the process, Laozi rejected the assumption made in other Chinese philosophies, including Confucianism, Mohism, and Legalism, that the *Way* is that of cultivation—controlling the self-regulating contexts of nature by deploying artificial machines and tools.
12. Hansen (1992) sees Daoists as perspectival relativists in contrast to the emphasis on rituals and rules of propriety by Confucians and utilitarianism by Mohists. This is still to place stress on social relations rather than the relations with nature that we propose as the original ground of controversy at the dawn of the agricultural revolution that emerged with intensive farming in China.
13. Trans. Chan (1963), p. 137.
14. Trans. Chan (1963), p. 139.
15. Ch. 2 of *Dao De Jing*. Trans. Chan (1963), p. 140. This chapter brings out more clearly the distinction between the way of cultivation and spontaneous generation.
16. This is not to deny that even in the Daoist tradition it later came to be interpreted metaphysically.
17. The concept *wu wei* in Daoism means non-doing or non-action. It is used by Laozi to show how when things are in harmony with the Dao their behavior is completely natural and effortless. Trees grow, stones fall to the ground, and birds build nests without contrivance and planning. Laozi held that the ideal of human behavior should also be without effort and striving. *Wu* may be translated as without and *Wei* as act or effort, so that *wu wei* becomes "without action" or "without effort". It later came to be incorporated in the apparently paradoxical notion of *wei wu wei*, that is, "action of non-action" or "action without action". For a study of its impact on early modern European economic thinking, see Gerlach (2005).
18. For an extended discussion, see Loy (1988), pp. 96–112.
19. See also Creel (1970: 54) quoting Wang Bi; Fung Yu-Lan (1966: 100–101). Wang Bi wrote a commentary on the *Dao De Jing*, and the text that accompanied it has traditionally been taken as the most authoritative until the Mawangdui Silk Texts were discovered in 1973. See Wing-tsit Chan (1963), especially Chapters 8, 63, 64, and 66 for these analogies.
20. However, adopting the naturalistic approach would require us to go beyond the non-dualist mystical path chosen by Loy—it requires us to embrace dualism by drawing a sharp distinction between human actions inspired by the masculine principle of dominating nature by cultivation, and those inspired by the feminine principle that yield to the spontaneous

actions of self-regulating nature. Loy is led to the non-dualist interpretation because he thinks that the Daoist paradoxes are, like those of Buddhist and Advaita Vedanta philosophies, intended to point us toward subject–object non-duality. (Loy 1988: 102–103). However, Loy's non-dual interpretation fails to account for the distinction between the Dao that can be described and the eternal but ineffable Dao that constitutes the opening line of the *Dao De Jing*.

21. Chapter 48. *Dao De Jing* trans. Chan (1963), p. 162.
22. Callicott gives an interpretation of the Biblical narrative of the expulsion from the Garden of Eden that supports this view. He argues that the story "read as the ethnohistory of cultures" traces the emergence of agricultural Neolithic humanity out of the Paleolithic state of nature. (Callicott 1994: 36). The same point has been made by Tucker and Grim who see it as referring to a period 12,000 years ago that saw the birth of plough agriculture. They describe the consequences:

    > It brought with it many human evils unknown to the earlier foragers, including slavery, patriarchy, organized warfare, and loss of ecological innocence. With the rise of farming, fields, and settlements began to displace animal habitats at alarming rates. As food sources became more stable, the human population began to increase dramatically, thus requiring still more habitat destruction. In many ways humans became not plain citizens of the planet but lords of the planet. Much of our dominion became domination. (Tucker and Grim 1994: 76–77)

23. Karyn Lai (2000) argues that we need to understand in a more nuanced way the positive evaluation of femininity, and the values associated with it, by the Daoists. She emphasizes, however, that the characteristic of feminine submissiveness connected with Daoism must be repudiated, although the Daoist notion of the complementarity of the masculine and feminine should be valued.
24. Carolyn Merchant writes:

    > Central to the organic theory was the identification of nature, especially the earth, with a nurturing mother: a kindly beneficent female who provided for the needs of mankind in an ordered, planned universe. But another opposing image of nature as female was also prevalent: wild and uncontrollable nature that could render violence, storms, droughts, and general chaos. Both were identified with the female sex and were projections of human perceptions onto the external world. The metaphor of the earth as a nurturing mother was gradually to vanish as a dominant image as the Scientific Revolution proceeded to mechanize and to rationalize the world view. The second image, nature as disorder, called forth an

> important modern idea, that of power over nature. Two new ideas, those of mechanism and of the dominion and mastery of nature, became core concepts of the modern world. (Merchant 1980: 2)

For similar ecofeminist critiques of modern science, and its reductionist mechanical paradigm and the values associated with it, which also take into account postcolonial perspectives, see Vandana Shiva and Maria Mies (1993) and Harding (1998).

25. Other feminists like Evelyn Fox Keller argue that Merchant's view is one sided and that Bacon's orientation is not that of a rapist but a seducer intent on making "a chaste and lawful marriage between mind and nature" (Keller 1985: 36).
26. Quoted in Merchant (1980), p. 168.
27. Cited in Merchant (1980), p. 170. See also Hess (1995), pp. 82–84.
28. Bacon's role in the scientific revolution came to be questioned only in the twentieth century after two eminent historians of science, Alexandre Koyre and E.J. Dijksterhuis, disparaged his contribution largely because he made no important scientific discoveries. When they question the importance of Bacon to science it is evident that Gross and Levitt judge him only in terms of his scientific contributions *per se*, which leads them to implicitly disparage his methodological endowment. This revaluation of Bacon reflects the unfortunate separation and specialization of contemporary science and philosophy, so that methodologists like Bacon are no longer taken to have had a significant impact on science [see Losee (2001: 56)].
29. Bacon, *Novum Organum*, Aphorism 3, p. 43.
30. ibid., Aphorism 41, p. 54.
31. ibid., Aphorisms 42–44, pp. 54–56.
32. ibid., Aphorism 68, p. 77.
33. ibid., Aphorism 98, pp. 107–108.
34. This Baconian view of science is now accepted by most scientists, and even affects our judgments regarding whether traditions of natural knowledge accumulated by other cultures could be deemed scientific. For example, there are those who deny that there was a premodern scientific tradition in China—Bodde argues that there was Chinese technology but no Chinese science (Bodde 1991: 358).
35. Much of the preceding discussion of the ecological processes in nature, that came to shape and be modified by agriculture, and their suppression and displacement by modern agro-industry, draws upon Capra (1982): 252–260. For a pioneering, albeit controversial, study of permaculture, see Mollison and Holmgren (1978).
36. The notion of exchange value has had a very long history from Aristotle to David Ricardo and Karl Marx. It was often distinguished from three other

major attributes of a commodity—use value, value, and price. However in contemporary neo-classical economics, which takes money-price to be sufficient to understand markets and trading practices, exchange value is no longer theorized explicitly. This had been rightly bemoaned by Alexander Gersch:

> In economics of all topics value is most disputed. This is because the theory of exchange which lies at the threshold of our science forms a connecting link between problems of a purely economic nature and the social. Moreover it acts as a point of departure for theoretical inferences affecting the entire domain of human economy. Its abstract nature renders an objective approach quite difficult and for all who have ventured to overcome these impediments it turned to be a stumbling block. Thus by its nature the theory of exchange value is a most ungrateful topic to be dealt with. Yet, being of essential importance to economics and a problem which was not solved, it invites adventurous minds to attempt its solution. (Gersch 1969: v)

37. Both of these factors were shaped by conceptions ultimately derived from the mechanical vision that inspired physics—a vision of production as mediated through machines and the market as a mechanism. For an illuminating study of the way conceptions in physics informed modern economic theory, see Mirowski (1989).
38. See John M. Hobson (2004); also Abu-Lughod (1989).
39. For a critical discussion of this orientation of classical economics, see Dugger, especially chapter 25 entitled "Instituted Process and Enabling Myth: The Two Faces of the Market" that critiques Adam Smith's natural law outlook. Dugger writes:

> The simple observation that the market is an instituted process rather than a natural equilibrium takes on great significance because it makes accountable men and women who exercise power behind the protection of the market myth. That simple observation eliminates their protection. Where the market is understood as an instituted process, those who institute it can be held responsible. (Dugger 1992: 235)

40. For a philosophical examination of the distinction between natural laws in physics and economic laws, see Clark (1992) who develops a sustained critique of Adam Smiths's natural law outlook.
41. Smith's study *An Inquiry into the Nature and Causes of the Wealth of Nations*, was originally published in 1776, in the same year as the American Revolution, and became a foundational text for classical economic theory.
42. For a discussion of these issues, refer to Adolfo Garcia de la Sienra (1992).

43. Actually what is commodified is not labor but labor power. For a critical discussion of this distinction, see Popper (1974), pp. 170–177.
44. The Dodo became extinct in the seventeenth century after its discovery in Mauritius. It is a flightless bird whose appearance is only now known through paintings and writings by observers shortly before its extinction.
45. In his *magnum opus*, *The Great Transformation*, published in 1944 Karl Polanyi describes the growth of the modern market economy and the modern state as interdependent and mutually reinforcing phenomena. He argues that the changes in social structure that promoted a competitive capitalist economy was brought about by the actions of a powerful modern state, which in turn came to enhance its power by liberating the forces of market capitalism. Both the forces of the modern state and the capitalist economy entailed the destruction of the social order that preceded them. In short, Polanyi argues that the free market economy and the ideology that inspired it were essentially outcomes of design and planning.
46. The 'means of production' refers to the physical inputs that are deployed in the production of economic goods and services, including tools, machines, and factories, as well as infrastructural and natural capital.
47. Before addressing the issue of how economic theory can learn from ecofeminist and Daoist epistemologies it is important to appreciate the different applications of the feminine principle in the two traditions. Ecofeminists appeal to the feminine principle to prevent destruction by the dominative orientation to nature inspired by the masculine principle of ecological contexts which support natural self-regulating powers. In contrast, Daoists appeal to the feminine principle to advise us to tap the self-regulating powers of nature by yielding to natural contexts rather than replacing them with artificial ones. One way of contrasting the two perspectives is by comparing Gaian and Daoist views.

    The Gaia theory is itself named after the Goddess who was revered as the supreme deity in pre-Hellenic Greece. She was worshipped in many cultures under various names and identified with the Earth as an animated being. According to Merchant, the view of the Earth as a living and spiritual being flourished in the Middle Ages and during the Renaissance, until it was displaced by the Enlightenment view of it as a mechanical system. Lovelock and Margolis named their Gaia hypothesis after the Earth Goddess precisely because they saw themselves as returning to the quasi-organic view of the earth held by many ancient cultures. Hence, when we compare the Daoist view inspired by the *yin* principle, with the Gaia theory named after the ancient Goddess of the earth, we are comparing two perspectives inspired by the feminine principle emanating at different times from opposite ends of the world—the ancient East and the modern West.

There is a remarkable convergence in the two viewpoints. This is not surprising because both are appealing to the feminine principle and self-regulating processes in nature. We have seen that the Gaian view can be described as quasi-organic, and the Daoist view as organic materialist.

Moreover, both Gaians and Daoists perceive agriculture as the greatest threat to human welfare. But their reasons for this shared view of agriculture are different. In the period of the Warring States, Daoists identify the attempt to cultivate nature as the root cause of the disruption of human society and the creation of gross inequalities of wealth and power. By contrast, Gaians see agriculture as a threat today because, by encouraging forest conversion into cultivable land, it is the primary cause of deforestation—especially in the tropical regions. This difference is clearly evident in the way Lovelock sees modern agriculture:

> There is no way for us to survive without agriculture, but there seems to be a vast difference between good and bad farming. Bad farming is probably the greatest threat to Gaia's health. We use close to 75 per cent of the fertile land of the temperate and tropical regions for agriculture. To my mind this is the largest and most irreversible geophysiological change that we have made. Could we use this land to feed us and yet sustain its climatic and chemical geophysiological roles? Could trees provide us with our needs and still serve to keep the tropics wet with rain? Could our crops serve to pump carbon dioxide as well as the natural ecosystems they replace? It should be possible but not without a drastic change of heart and habits. I wonder if our great-grandchildren will be vegetarian and if cattle will live only in zoos and in tame life parks. (Lovelock 1989: 179)

Finally, Daoists and Gaians emphasize the significance of self-regulating natural technologies in furnishing human beings with the resources needed to meet their basic needs. The Daoists see these technologies as important for the production and provision of food and raw materials; the Gaians in providing us with clean water, air to breathe, and moderate temperatures. Just as the ancient Daoists warn of the dangers of interfering with nature's processes by using mechanical technologies, so that we become increasingly dependent upon them with attendant dire social and economic consequences, so do modern Gaian ecofeminists warn us of the possibility of such dependency, because it may threaten not just our basic needs of food, clothing, and shelter but also our more urgent needs of fresh water, clean air, and livable climates.

Lovelock also argues that if we pursue present destructive relations with the biosphere then we may be forced to take over the maintenance of such life support systems as clean water and air, and livable climates, after we have undermined nature's self-regulating powers:

> This could happen if, at some intolerable population density, man had encroached upon Gaia's functional power to such an extent that he disabled her. He would wake up one day to find that he had the permanent lifelong job of planetary maintenance engineer. Gaia would have retreated into the muds, and the ceaseless intricate task of keeping all of the global cycles in balance would be ours. (Lovelock 1982: 132)

Should such a situation arise then it would be a second expulsion from the Garden of Eden—one more painful and toilsome than the first. Moreover it may bring a new period of Warring States, this time on a global scale, as nations and tribes struggle for scarce resources of clean air, drinkable water, agricultural land, and fisheries to meet basic needs now taken for granted by many.

However, despite their similarities there is a significant difference between the Gaian and Daoist perspectives. The Gaians warn us to limit our unrestrained exploitation of nature so that the healthy functioning of the biosphere would not be undermined. In short humans are encouraged to conduct their economic activities within the confines of the biosphere's ecological constraints. By contrast, the Daoists do not perceive nature as threatened; instead they see only humans as threatened by their attempts to control nature and interfere with its processes. They consider that there would be an ample supply of food and resources provided people are prepared to adopt a simple lifestyle.

The significant difference is that Gaians warn us against an excessively dominative orientation to nature that could destroy its self-regulating powers, but Daoists teach us to use these powers in order to produce goods and services for ourselves. Clearly the Daoist notion of managing production more by natural technologies than human technologies is still relevant today—even if it is not taken as practical to apply it to agriculture. This is because natural technologies continue to play an important role in delivering essential goods we need—clean air, pure water, and moderate temperatures at the planetary level. If we degrade the ecological systems that now provide them without our intervention even such functions might have to be engineered in the future. At that point spaceship Earth would truly have become a mechanical spaceship in which life-support functions would have to be maintained with great effort and energy by human, rather than natural, technologies.

48. *Small Is Beautiful: Economics As If People Mattered* was first published in 1974 and became one of the most influential works in both developmental and environmental circles. For an update of his views, see Schumacher (1999).

49. Such an approach has motivated a number of environmental philosophers to extend various ethical orientations—Rawlsian ethics, stewardship ethics, utilitarian ethics, or intrinsic value theories—to embrace future generations as objects of present moral concern. See Partridge (1981).
50. See Simon and Kahn (1984) for an extended argument along these lines.
51. For a comprehensive review of the connection between artificial selection and evolutionary theory, see Bajema (1982). According to Briggs and Peat (1984: 189) "Scientists have never seen an entirely new species created, though the evolution of many new varieties have been created by animal and plant breeders."
52. The role of forests in maintaining global human welfare is recognized in the now entrenched practice of "debt-for-nature" swaps in which developing countries exchange their external debts for funds dedicated to environmental protection—especially the protection of tropical rainforests. See Fuller (1988).
53. This accords with the intuition of many environmental philosophers that natural systems have an intrinsic value, and that one cannot place an economic value on ecosystems or species (Devall and Sessions 1985: 115–118). Since a forest is impossible to reproduce once it has been destroyed—its cost of reproduction is indefinitely large—the forest, as a whole, would have to be treated as having an indefinitely large economic value. The same could be said of any ecosystem or species which is irretrievably destroyed—the economic value lost is immeasurable.

    However, provided an ecosystem or species is not degraded too far, so that it continues to retain its self-reproductive potential, the cost of restoring what is lost is finite. This again accords with our intuition that human beings can treat ecosystems or species as economic resources provided an already prior commitment has been made to use them in a sustainable fashion. This is stressed in the *World Commission on Environment and Development Report* which writes:

    > The process of economic development must be more soundly based upon the realities of the stock of capital that sustains it. This is rarely done in developed or developing countries. For example, income from forestry operations is conventionally measured in terms of the value of timber and other products extracted, minus the cost of extraction. The costs of regenerating the forest is not taken into account, unless money is actually spent on such work. (WCED 1987: 52)

54. By contrast, for human artifacts like factories, the cost of production and the cost of reproduction are both the same and of finite value (assuming that we have adjusted for inflation over time of costs and wages). The reason for this difference is evident: the cost is measured in terms of

anthropocentric parameters, but since natural systems are self-regulating and reproduce themselves without human intervention or effort, they produce themselves at no human cost, although they cannot be reproduced, once lost, at any human cost.

However, this does not mean that we need two laws of exchange value; that the economic supply-demand law has to be complemented with another exchange law. Actually the process of establishing ecological boundary conditions itself changes the pattern of supply and demand; it places constraints on the supply and has effects on the pattern of demand. The same supply-demand law yields different equilibrated prices under different boundary conditions. The problem is not with the law of supply and demand but with the auxiliary assumption that informs much of modern economic science—that natural technologies are similar to mechanical technologies; that they can be assimilated into the latter; that the universe of nature is only an extension of the clockwork universe.

In fact, placing boundary conditions on the operation of the supply-demand law is not new to economic practice. Economic science itself has always, even in the modern era, only allowed the law of supply and demand to operate within boundary conditions. Such boundary constraints allow us to sell our labor, but not ourselves: slavery is prohibited. There are also constraints on the sale of harmful and addictive drugs regardless of demand and supply; tariffs that alter prices of commodities as they enter nations; subsidies for agriculture that affect prices of certain basic foods and so on. Hence, the exchange law has always operated within the framework of politically defined boundary conditions. Establishing ecologically inspired boundary conditions is but an extension of this process. It allows us to protect natural capital and give it economic value by creating new boundary conditions—conditions based upon criteria of sustainability for the operation of the law of supply and demand. As a result the operation of the law is made ecosensitive.

The need for boundary conditions arises because of the inseparability of ecological and economic concerns. Just as political economy widens the notion of economics beyond the narrow concern with mathematical models assumed to operate acontextually, and locates them within the larger sociopolitical context within which they actually function, so would an ecological economics recognize itself as located within a wider framework of production.

Such an ecoeconomics would recognize the following: that production of goods and services for human beings is managed by both natural and humanly instituted systems; that sensitivity to the former as well as the latter requires us to consider both the context of production by nature and by human beings; and that ignoring nature's context by merely concerning ourselves with the human distorts our understanding of the processes of production and distribution that is the focus of concern of economic science.

The goal of economics to become an acontextual science, one that ignores the role of natural habitats in shaping the processes it studies, is inspired by a mechanical philosophy—the same philosophy that failed to recognize physical properties as arising in the context of an experimental arrangement, or perceptual properties as rooted in a theoretical context. An acontextual economic science sees prices as established without boundary conditions—a goal it can pursue only by ignoring nature's role in production, and one, moreover, it deludes itself as pursuing because in reality there are politically dictated boundary conditions that always constrain the processes it studies.

55. See Hawken et al. (1999).

Bibliography

Abu-Lughod, Janet L. 1989. *Before European Hegemony: The World System A.D. 1250–1350*. Oxford: Oxford University Press.

Bacon, Francis. 1561–1626. *Novum Organum*. Translated and Edited by Peter Urbach & John Gibson (1994). Chicago: Open Court.

Bajema, Carl Jay, ed. 1982. *Artificial Selection and the Development of Evolutionary Theory*. Pennsylvania: Hutchinson Ross Publishing Co.

Bala, Arun. 2006. *The Dialogue of Civilizations in the Birth of Modern Science*. New York: Palgrave Macmillan.

Bodde, Derek. 1991. *Chinese Thought, Society and Science: The Intellectual and Social Background of Science and Technology in Pre-modern China*. Honolulu: University of Hawaii Press.

Briggs, John P., and David Peat. 1984. *Looking Glass Universe: The Emerging Science of Wholeness*. New York: Simon and Schuster.

Butlin, J.A., ed. 1981. *Economics and Resources Policy*. London: Longman.

Callicott, J. Baird. 1994. *Earth's Insights: A Survey of Ecological Ethics from the Mediterranean Basin to the Australian Outback*. Berkeley: University of California Press.

Capra, Fritjof. 1982. *The Turning Point: Science, Society, and the Rising Culture*. New York: Simon and Schuster.

Chan, Wing-Tsit. 1963. *A Source Book in Chinese Philosophy*. Princeton, NJ: Princeton University Press.

Clark, Charles Michael Andres. 1992. *Economic Theory and Natural Philosophy: The Search for Natural Laws of Economy*. Cheltenham: Edward Elgar Publishing.

Creel, Herlee G. 1953. *Chinese Thought: From Confucius to Mao Tse-Tung*. Chicago: University of Chicago Press.

Creel, Herlee G 1970. *What is Taoism?* Chicago: University of Chicago Press.

De Klemm. 1985. Culture and Conservation: Some Thoughts for the Future. In ed. McNeely and Pitt, 1985: 245–246.

Devall, Bill, and George Sessions. 1985. *Deep Ecology*. Salt Lake City: Peregrine Smith Books.

Dugger, William M. 1992. *Underground Economics: A Decade of Institutionalist Dissent*. New York: Routledge.

Dyke, C. 1981. *Philosophy of Economics*. Englewood Cliffs, NJ: Prentice Hall.

Fuller, Kathryn. 1988. *Debt-for-Nature Swaps: A New Conservation Tool*. USA, in World Wildlife Fund Letter, World Wildlife Fund.

Fung, Yu-Lan. 1966. *A Short History of Chinese Philosophy*. New York: Free Press.

Garcia de la Sienra, Adolfo. 1992. *The Logical Foundations of the Marxian Theory of Value*. Netherlands: Springer.

Gerlach, Christian. 2005. *Wu-Wei in Europe: A Study of Eurasian Economic Thought*. London School of Economics. Working Paper No. 12/05. London: Department of Economic History, London School of Economics.

Gersch, Alexander. 1969, 1972. *On the Theory of Exchange Value Volumes 1 & 2*. Berkeley, CA: University of California.

Gross, Paul R., and Norman Levitt. 1994. *Higher Superstition: The Academic Left and Its Quarrels with Science*. Baltimore, MD: The Johns Hopkins University Press.

Harding, Sandra. 1998. *Is Science Multicultural? Postcolonialisms, Feminisms, and Epistemologies*. Bloomington, IN: Indiana University Press.

Hansen, Chad. 1992. *A Daoist Theory of Chinese Thought: A Philosophical Interpretation*. New York: Oxford University Press.

Hawken, Paul, Amory Lovins, and Hunter Lovins. 1999. *Natural Capitalism: Creating the Next Industrial Revolution*. Boston: Little, Brown and Company.

Herschel, John F.W. 1851. *A Preliminary Discourse on the Study of Natural Philosophy*. London: Longman, Brown, Green & Longmans.

Hess, David J. 1995. *Science and Technology in a Multicultural World: The Cultural Politics of Facts and Artefacts*. New York: Columbia University Press.

Hobson, John M. 2004. *The Eastern Origins of Western Civilisation*. New York: Cambridge University Press.

Jones, W.T. 1969. *The Classical Mind: A History of Western Philosophy*. New York: Harcourt Brace & World.

Keller, Evelyn Fox. 1985. *Reflections on Gender and Science*. New Haven: Yale University Press.

Kohn, Livia, and Michael LaFargue, eds. 1998. *Lao-Tzu and the Tao Te-Ching*. Albany: State University of New York Press.

Lai, Karyn. 2000. The Daodejing: Resources for Contemporary Feminist Thinking. *Journal of Chinese Philosophy* 27(2): 131–153.

Lloyd, Geoffrey, and Nathan Sivin. 2002. *The Way and the World: Science and Medicine in Early China and Greece*. New Haven: Yale University Press.
Losee, John. 2001. *A Historical Introduction to the Philosophy of Science*. USA: Oxford University Press.
Lovelock, James. 1982. *Gaia: A New Look at Life on Earth*. Oxford: Oxford University Press.
———. 1989. *The Ages of Gaia: A Biography of our Living Earth*. Oxford: Oxford University Press.
Loy, David. 1988. *Nonduality: A Study in Comparative Philosophy*. New Haven & London: Yale University Press.
McNeely, Jeffrey, and David Pitt. 1985. *Culture and Conservation: The Human Dimension of Environmental Planning*. London: Croom Helm Ltd.
Merchant, Carolyn. 1980. *The Death of Nature: Women, Ecology, and the Scientific Revolution*. New York: Harper & Row.
———. 1992. *Radical Ecology: The Search for a Livable World*. New York: Routledge.
Mies, Maria, and Vandana Shiva. 1993. *Ecofeminism*. Halifax, NS: Fernwood Publications.
Mirowski, Philip. 1989. *More Heat Than Light: Economics as Social Physics, Physics as Nature's Economy*. Cambridge, New York: Cambridge University Press.
Mollison, Bill, and David Holmgren. 1978. *Permaculture One: A Perennial Agriculture for Human Settlements*. Australia: Transworld Publishers.
Needham, Joseph. 1956. *Science and Civilization in China*. Cambridge, MA: Cambridge University Press.
Partridge, Ernest, ed. 1981. *Responsibilities to Future Generations: Environmental Ethics*. New York: Promethean Books.
Peerenboom, R.P 1991. Beyond Naturalism: A Reconstruction of Daoist Environmental Ethics. *Environmental Ethics* 13(1): 3–22.
Polanyi, Karl. 1977. *The Livelihood of Man*. New York: Academic Press.
Popper, Karl. 1974. *The Open Society and Its Enemies: Hegel and Marx*, vol 2. London: Routledge & Kegan Paul.
Robinson, Joan. 1973. *Economic Philosophy*. Middlesex, England: Penguin Books.
Smith, Adam. 1961. *The Wealth of Nations*. With an introduction, notes, marginal summary and an enlarged index by Edwin Cannan. England: Methuen & Co. Ltd. v.1. Originally published in 1776.
Schumacher, E.F. 1974. *Small is Beautiful: A Study of Economics as if People Mattered*. London: Abacus.
Schumacher, E.F. (1999) *Small Is Beautiful, 25th Anniversary Edition: Economics As If People Mattered: 25 Years Later . . . With Commentaries*. Dublin: Hartley and Marks Publishers.

Shankman, Steven, and Stephen W. Durrant. 2002. *Early China, Ancient Greece: Thinking Through Comparisons.* Albany, NY: State University of New York Press.

Simon, Julian, and Herman Kahn, eds. 1984. *The Resourceful Earth: A Response to Global 2000.* Oxford: Basil Blackwell.

Toulmin, Stephen. 1990. *Cosmopolis: The Hidden Agenda of Modernity.* New York: The Free Press.

Tucker, Mary Evelyn, and John A. Grim, eds. 1994. *World Views and Ecology: Religion, Philosophy, and the Environment.* Maryknoll, NY: Orbis Books.

World Commission on Environment and Development Report. 1987. *Our Common Future.* New York: Oxford University Press.

CHAPTER 5

Complementarity and Unity of Knowledge

We began this study with Bohr's call to enrich our understanding of the epistemological discoveries of quantum physics by turning to similar discoveries made by Eastern thinkers. The process of explaining how such parallel developments of thought came about not only vindicates Bohr's view that complementarity has significant applications beyond physics in disciplines as diverse as biology, psychology, and the social sciences but also his claim that parallel notions exist in Eastern thought. However, we cannot conclude this study without also investigating why Bohr's forays into epistemology have been for so long ignored or disparaged by not only historians and philosophers of science, but even scientists themselves. Moreover, even the few historians who have paid attention to his efforts to extend complementarity beyond physics, such as Folse and Pais, hardly mention his interest in Indian or Chinese philosophical traditions. Indeed when Bohr's interest in such intercultural dialogue does get noticed, as is the case with Beller and Sokal, it is only invoked in order to parody and discredit his views concerning complementarity.

However, it would be too one-sided to attribute the general reluctance to pay more serious attention to Bohr's efforts to link with Eastern thought to simple-minded prejudice alone. Bohr himself may have given grounds for the suspicion that even if complementarity lends itself to extension into areas of science beyond physics, its links with Eastern philosophies are still likely to be problematic. Indeed the explanation Bohr gives for why complementarity can be extended beyond physics may also have led Folse and Pais to reasonably infer that Indian and Chinese epistemological

Arun Bala, *Complementarity Beyond Physics*,
DOI 10.1007/978-3-319-39784-9_5

notions may not illuminate complementarity. Consider Folse's account of Bohr's grounds for extending complementarity as a general framework for science:

> The key to Bohr's extension of complementarity beyond atomic physics is the fact that in description of certain phenomena the required observational interaction has this indivisible quality. From this situation it follows that the interaction is "uncontrollable", thus forcing us to regard the representation of the object as isolated from observational interaction as an abstraction. This conclusion in turn implies that a proper understanding of science must prohibit regarding the terms in which such an abstraction is expressed as corresponding to the properties of an independent reality. Bohr believed that he had discovered violations of precisely this prohibition in controversies where rival descriptions of some range of phenomena were interpreted as disputes about the nature of independently real objects. Specifically, he found such situations in the "*free will/determinist*" *controversy*, which he interpreted as a dispute in psychology, and the "*mechanist/vitalist*" *controversy* in biology. In both cases he believed that the resolution of the dispute, like the case of wave-particle dualism, would not involve a victory for one side or the other … He also tended to see a similar situation in "*nature or nurture*" *disputes* in cultural anthropology. [My emphasis] (Folse 1985: 174)[1]

Folse rightly argues that Bohr's motivation for extending complementarity into biology, psychology, and the social sciences is an effort to generalize an epistemological discovery he made in atomic physics. This is the recognition that every act of observation involves two stages—first, arbitrarily dividing a unitary whole into an observing and an observed part, and second, using an "uncontrollable interaction" of the observing system on the observed system, to make the observation. In psychology, we have to divide the unitary psyche into a part doing the observing and a part being observed; in biology, the unitary organism into a part observed from the rest of the organism as the context in which it gets observed; and in anthropology, the unitary culture formed by the embedded anthropologist into an observing scientist and the rest of the culture observed. Then we have to use an interaction of one part on the other to arrive at an observed result.

However, the complexity of the processes involved between the interacting systems make the interactions uncontrollable and unpredictable. Thus, the introspecting part of the subject uncontrollably influences the observed psychological event, for example, observing our anger changes it because we are no longer carried away by it. Observing a cell requires isolating it and uncontrollably influencing the processes that would have taken place in it within its location in the organism without our intervention. The anthropologist in a culture also influences it uncontrollably in the process of studying it. Bohr suggests that it is the influence of the act of observation in uncontrollably shaping the observed result that makes the epistemological discovery, which is embodied in the complementarity framework, have wider general applicability in science.

However, Bohr also gave up his appeal to uncontrollable interactions largely as a result of his debates with Einstein concerning the interpretation of quantum theory. After 1935, when he wrote his response paper to Einstein, he explained complementarity not by appeal to uncontrollable interactions but to the intrinsic wholeness in atomic processes following the discovery of the quantum of action. However, even this account fails to show why complementarity should appear in systems where the action is large so that its quantum effects can be ignored as in classical physics. This should lead us to suspect that complementarity epistemology can in general be ignored in the macro-world of psychology, biology, and anthropology and also Eastern thought that had no access to the micro-world of quantum physics.

Another reason why Bohr's call to extend complementarity beyond physics has not won general acceptance is that we do not find any indeterminacy that permits only statistical predictions in the disciplines he invokes. There is no uncertainty in saying that in the duck-rabbit configuration we will see an 'ear' when we adopt the context of seeing the rabbit, or that we will find an 'eye-cell' grow when we find an embryonic cell in an eye context. The property arises in dependence on the context, but not in an indeterminate fashion that makes it impossible to predict the outcome.

But the situation is more complex in the case of observation of a quantum property. Even after we choose to measure the momentum of a particle, say, there continues to remain uncertainty about the precise value of the observed momentum. We can predict the qualitative nature of the property measured—namely, that it will be momentum—but not its quantitative measure. Thus uncertainty only pertains to its quantitative value, but not its qualitative outcome. Similarly when we choose to measure the spin of a particle along the x-axis we can predict that it will qualitatively be an x-spin, but remain uncertain about whether it will be up or down.

This shows that it is important to separate two different issues in quantum physics. First, there is the issue of whether the observed object can be isolated from the observing system. Second is the issue of whether the outcome of the influence of the observing system is indeterminate. In the case of quantum physics, as well as gestalt psychology and biology, the observed system and observing system cannot be separated because the property arises in the observational context. However, what originates in this context can be precisely predicted in advance in all cases except one. The exception is the quantum case where we can only predict the qualitative nature of the property that will grow in advance, but not its precise quantitative value. Hence, it is only in the quantum case that the indeterminacy arising from the finiteness of the quantum of action makes possible only probabilistic prediction of the possible results of observation.

Bohr fails to conceptually distinguish the influence that requires the complementarity perspective from the influence that requires us to make probabilistic predictions in quantum physics. The former is grounded in the fact that the observed system cannot be separated from its context of observation, but probabilistic predictions are required only when the interactions involved make predictions uncertain. However, by using the notion of the finiteness of the quantum of action to explain the complementarity viewpoint, Bohr links complementarity to Heisenberg's uncertainty principle. It is this close association of complementarity and the uncertainty principle that is used in the Copenhagen interpretation of quantum theory. However, such a connection also comes at a price—it suggests that Bohr cannot be right in thinking that we can learn from Indian and Chinese epistemological views, since nothing like the uncertainty principle or probabilism has been formulated in their epistemological frameworks. This has made it difficult for others to go along with his view that Eastern philosophies can illuminate issues in quantum epistemology.

Surprisingly Folse himself notes that the complementarity viewpoint can actually be separated from the Copenhagen interpretation, and even goes so far as to suggest that the failure to do so became an important factor that obstructed Bohr's efforts to promote complementarity as a general framework for science:

> [U]nfortunately history has not been altogether kind to his [i.e., Bohr's] philosophical endeavors. Instead of being understood as a general framework within which the new physics was to be justified as an objective description of nature, complementarity came to be identified with the so-called "Copenhagen Interpretation" of quantum theory. Furthermore this Copenhagen Interpretation, and complementarity with it, came to be com-

monly associated with the writings of a whole group of physicists, who may not always have fully grasped what Bohr was saying. [Folse (1985): 6]

Although Folse argues in the above passage for the need to liberate complementarity from the Copenhagen view, because the latter assimilates extraneous notions alien to Bohr's position, he fails to see the importance of separating complementarity from its association with the uncertainty principle and quantum indeterminacy. However, by tracing it to grown properties in nature our study has liberated complementarity from its association with the Copenhagen viewpoint. We have shown that complementarity arises in the quantum situation because measurement is not merely identifying a pre-existing property of the observed system—it is growing the property measured. By contrast, uncertainty arises at the same time because of the finiteness of the quantum of action, which is the basis for the indeterminacy principle. Although the Copenhagen viewpoint combines complementarity and indeterminacy, the roots of complementarity do not lie in the uncertainty principle. Hence, we can have features of complementarity even in situations where we do not have features of indeterminacy.

Another reason why Bohr's attempt to extend complementarity beyond physics failed is that he did not give any underlying unifying explanation for why such an effort makes sense. Although he spoke in many places of complementarity as a framework for promoting 'unity of knowledge', bringing together 'the method of analysis and synthesis', and generalizing the principle of causality that informed classical mechanical science, he did not offer any explanation for why such an alternative framework of epistemology to causal mechanism was needed. This becomes evident in a talk he gave to the *Second International Conference for the Unity of Science* in 1936 on 'Causality and Complementarity'. He said:

> On several occasions I have pointed out that the lesson taught us by recent developments in physics regarding the necessity of constant extension of the frame of concepts appropriate for the classification of new experiences leads us to a general epistemological attitude which might help us to avoid apparent conceptual difficulties in other fields as well ... [It] presents us with a situation concerning the analysis and synthesis of experience which is entirely new in physics and forces us to replace the ideal of causality by a more general viewpoint usually termed "complementarity." [Bohr (1999): 39–41]

He makes the same point nearly 20 years later in an address delivered in 1955 at the opening session of the United Nations conference entitled 'Physical Science and Man's Position' without also giving any reason for why complementarity would find applicability beyond physics:

> The importance of the epistemological lesson which the exploration of the world of atoms has given us must be seen in the background of the impact of the mechanical conception of nature on general thinking through the centuries. Above all, the recognition of an inherent limitation in the scope of the deterministic description within a field of experience concerned with fundamental properties of matter, stimulates the search in other domains of knowledge for similar situations in which the mutually exclusive application of concepts, each indispensable in a full account of experience, calls for a complementary mode of description. [Bohr (1999): 104]

In this study we have attempted to show that complementarity is a response to grown properties in nature. We have already seen that complementarity in biology arises because things grow by virtue of an interaction of a template genetic code with the context of its environment—it is rooted in the *contextual origination* of grown properties. Similarly psychological complementarity arises because we see the world in terms of gestalts that originate when a template configuration is read through a theoretical framework. This *contextual identification* of properties itself is grounded in the way our senses have evolved to identify things in their contexts since grown properties arise in response to their environments. The anthropological complementarity that Bohr refers to is rooted in both the natural template and cultural context interactions when we engage with the world both in constituting ourselves and in shaping nature. Here it is the *contextual regulation* of properties that is significant. The general interaction of template and context in all these cases can be traced back to grown properties in nature.

In the mechanistic analytic framework it is generally assumed that observed properties are determined by the template and that the context plays no role in crucially forming them. The phenotype properties are determined completely by the genotype—the environment only furnishes the raw material for the growth of the phenotype. Perceptual experience is completely determined by the sensory stimuli from the template of the configuration and not at all shaped by our theoretical beliefs. The traits of human personalities observed are influenced by biological inheritance, and culture has only a marginal role in their formation. Hence, it is possible to see a completely causal link between the template—genetic code, configuration, and inherited human nature—and the articulated phenotype, percept, and human personality.

However, recognition of grown properties requires us to see phenotypes, percepts, and personalities as underdetermined by the template, and to see that a complete understanding of any observed trait in the phenotype, percept, and personality is also shaped by the corresponding context. It is the template–context interaction that explains the properties observed.

Yet, explaining a property in terms of the template and in terms of the context seem mutually exclusive. There have been ongoing controversies among biologists whether genotype or environment is the fundamental influence on phenotype properties; among psychologists whether sensory stimuli or theoretical contexts have the greatest influence on perceptual experience, and among social scientists whether nature or nurture has the largest impact on human personality.

The template–context duomorphism also explains why there are paradoxes in Eastern thought—in Buddhism and Daoism in particular as noted by Bohr in the beginning of this study—that seem to violate standard laws of logic in the way quantum properties do. We have traced these to the context dependence of properties observed and the need to take into account mutually exclusive contexts to fully explain these properties. Such contexts arise when we think of the measuring contexts for quantum properties, the theoretical contexts that determine how we perceive a gestalt configuration, the environmental contexts for biological properties, and the social contexts that influence properties in humans and cultivated nature.

Moreover, we found that this also suggests that both Buddhist and Daoist epistemologies can, in their different ways, enrich contemporary philosophy of the natural and social sciences in ways that Bohr did not imagine. First theory-ladenness of observations, that has become such a critical concern in post-positivist and post-empiricist philosophy of science, can profit from the insights of Indian philosophy, especially by taking into account deautomatizing technologies as ways of freeing our sensibilities from consolidated but outdated theoretical perspectives, and informing them with newer, more powerful frameworks. Second, the role of self-regulating processes in nature suggest that economic theory—and philosophy of the social sciences, in general—can learn from Daoist epistemology how to inform our interactions with nature by yielding to nature's self-regulating processes. In a fashion it takes us beyond the

Baconian dictum—namely that we must obey the laws of nature in order to command natural contexts—one step further. Daoist epistemology suggests that we must also obey natural contexts in order to live in harmony with nature. This is what the paradoxes of action, knowledge, and morality in Daoism are all about.

Thus the notion of grown properties gives a new framework for understanding complementarity that liberates it, as Bohr intended, to serve as an objectivist epistemological framework for the natural, biological, psychological, and social sciences, and for forging links between Western and Eastern traditions of epistemology. It does this by not only dissociating complementarity from the Copenhagen interpretation, which suggests that it cannot have relevance for the macro-world where the effects of the quantization of action can be ignored, but also by showing why complementarity is necessary as a framework for going beyond the reductive notion of causality in the mechanical view of the cosmos. Moreover it reveals why complementarity epistemology can profit by learning from Eastern epistemologies to enrich contemporary philosophy of the natural and social sciences. In this regard, the current study vindicates Bohr's lifelong efforts to elevate complementarity as a general epistemological principle to replace causality following the demise of classical physics and the rise of quantum theory.

Note

1. Folse goes on to quote Bohr:

> I am far from sharing, however, the widespread opinion that the recent development in the field of atomic physics could directly help us in deciding such questions as "mechanism or vitalism" and "free will or causal necessity" in favor of the one or the other alternative. Just the fact that the paradoxes of atomic physics could be solved not by a one-sided attitude towards the older problem of "determinism or indeterminism", but only by examining the possibilities of observation and definition, should rather stimulate us to a renewed examination of the position in this respect in the biological and psychological problems at issue. (Bohr quoted in Folse (1985: 174)) Original in Bohr (1937: 295).

Bibliography

Bohr, Niels. 1937. Causality and Complementarity. *Philosophy of Science* 4(3): 289–298.

———. 1999. *Collected Works Vol. 10: Complementarity Beyond Physics (1928–1962)*, ed. David Favrholdt. Elsevier.

Folse, Henry J. 1985. *The Philosophy of Niels Bohr: The Framework of Complementarity*. Amsterdam: Elsevier Science Publishers.

Name Index

Note: Page numbers followed by 'n' refer to notes.

Arun Bala, *Complementarity Beyond Physics*,
DOI 10.1007/978-3-319-39784-9

Y

Z

Subject Index

Note: Page numbers followed by 'n' refer to notes.

Arun Bala, *Complementarity Beyond Physics*,
DOI 10.1007/978-3-319-39784-9